Foucault and International Relations

The recent debate about biopolitics in International Relations (IR) theory may well prove to be one of the most provocative and rewarding engagements with the concept of power in the history of the discipline. Building on Foucault's arguments concerning the role played by the concept of security in 19th-century liberal government, numerous IR scholars are now arguing for the relevance of his theories of biopolitics and governmentality for understanding the Global War on Terror (GWOT) and broader issues of security and governance in the post 9/11 world.

Conversely, others have criticized this idea. Marxist and Communitarian scholars have challenged the notion that the category of biopolitics can be 'scaled' up to the level of international relations with any analytical precision. This edited volume covers these debates in IR with a series of critical engagements with Foucault's own thought and its increasing relevance for understanding international relations in the post 9/11 world.

This book was based on a special issue of *Global Society.*

Nicholas J. Kiersey is Assistant Professor of Political Science at Ohio University-Chillicothe. He has published research on "world state" theory, scale and bio-politics in the War on Terror, and the European Union's attitude to Turkish accession. His current research focuses on discourses of neoliberal capitalist subjectivity and the "debate about empire" in IR theory.

Doug Stokes is a Senior Lecturer in International Relations at the University of Kent, Canterbury. His work covers critical international relations theory and US foreign policy. His most recent book is called *American Hegemony and Global Energy Security* and is due out in 2010 with the Johns Hopkins University Press.

Interventions

Edited by: Jenny Edkins, Aberystwyth University and Nick Vaughan-Williams, University of Exeter

'As Michel Foucault has famously stated, "knowledge is not made for understanding; it is made for cutting." In this spirit The Edkins - Vaughan-Williams Interventions series solicits cutting edge, critical works that challenge mainstream understandings in international relations. It is the best place to contribute post disciplinary works that think rather than merely recognize and affirm the world recycled in IR's traditional geopolitical imaginary'

Michael J. Shapiro, University of Hawai'i at Mãnoa, USA

The series aims to advance understanding of the key areas in which scholars working within broad critical post-structural and post-colonial traditions have chosen to make their interventions, and to present innovative analyses of important topics.

Titles in the series engage with critical thinkers in philosophy, sociology, politics and other disciplines and provide situated historical, empirical and textual studies in international politics.

Critical Theorists and International Relations
Edited by Jenny Edkins and Nick Vaughan-Williams

Ethics as Foreign Policy: Britain, the EU and the Other
Dan Bulley

Universality, Ethics and International Relations: A Grammatical Reading
Véronique Pin-Fat

The Time of the City: Politics, Philosophy, and Genre
Michael J. Shapiro

Governing Sustainable Development: Partnership, Protest and Power at the World Summit
Carl Death

Insuring Security: Biopolitics, Security & Risk
Luis Lobo-Guerrero

Foucault and International Relations
New Critical Engagements
Edited by Nicholas J. Kiersey and Doug Stokes

International Relations and Non-Western Thought
Imperialism, Colonialism and Investigations of Global Modernity
Edited by Robbie Shilliam

Foucault and International Relations

New Critical Engagements

Edited by

Nicholas J. Kiersey and Doug Stokes

Routledge
Taylor & Francis Group
LONDON AND NEW YORK

First published 2011
by Routledge
2 Park Square, Milton Park, Abingdon, Oxon, OX14 4RN

Simultaneously published in the USA and Canada
by Routledge
270 Madison Avenue, New York, NY 10016

Routledge is an imprint of the Taylor & Francis Group, an informa business

This book is a reproduction of *Global Society*, vol. 23, issue 4. The Publisher requests to those authors who may be citing this book to state, also, the bibliographical details of the special issue on which the book was based

Typeset in Times New Roman by Taylor & Francis Books

British Library Cataloguing in Publication Data
A catalogue record for this book is available from the British Library

ISBN13: 978-0-415-57983-4

Contents

Notes on contributors

Leonie Ansems De Vries is a PhD candidate in the Department of War Studies at King's College London. Via the concept of the milieu and the notion of politics as movement, her thesis seeks to produce an alternative account of political life in relation to, and in the context of, (dis)order(ing).

Jonathan Joseph is Reader in Politics and International Relations at the University of Kent, Canterbury. His books include *Hegemony: A Realist Analysis* (Routledge, 2002), *Marxism and Social Theory* (Palgrave, 2006) and *Realism, Philosophy and Social Science* (with Kathryn Dean, John Roberts and Colin Wight, Palgrave, 2006). His most recent book is *Scientific Realism and International Relations* (edited with Colin Wight, Palgrave, 2010). He is currently working on a book on governmentality and International Relations.

Nicholas J. Kiersey is Assistant Professor of Political Science at Ohio University—Chillicothe. He has published research on "world state" theory, scale and biopolitics in the War on Terror, and the European Union's attitude to Turkish accession. He is currently writing a book on economic governmentality, neoliberal capitalist subjectivity and the global financial crisis.

Halvard Leira is Research Fellow at the Norwegian Institute of International Affairs (NUPI). His research interests converge on international political thought in a broad sense and, in particular, on foreign policy, diplomacy and sovereignty. He has a penchant for obscure themes and even more obscure sources. He has published extensively in Norwegian, most notably the centenary history of the Norwegian foreign service (with Iver B. Neumann), and in *Review of International Studies*, *Leiden Journal of International Law*, *Hague Journal of Diplomacy* and *Cooperation & Conflict*.

Ivan Manokha is Lecturer in International Relations and Vice-Dean of the Graduate School at the Institut d'Etudes Politiques (Sciences Po) of Paris. He completed his doctoral thesis on the political economy of humanitarian intervention at the University of Sussex in 2004. He has taught Human Rights at the London School of Economics and various International Relations courses at the University of Sussex, the University of Le Havre, and the ESSEC Business School in Paris. His research interests include historical development of the concept of human rights and its relationship with capitalism, Just War Theory, humanitarian intervention in the late modern global political economy, and business ethics.

Andrew W. Neal is a lecturer in Politics and International Relations at the University of Edinburgh. He is interested in the politics of security, critical methodologies, Continental philosophy and constitutional law. He is the author of *Exceptionalism and the Politics of Counter-terrorism: Liberty, Security and theWar on Terror* (Routledge, 2009) and editor (with Michael Dillon) of *Foucault on Politics, Security and War* (Palgrave, 2008).

Doerthe Rosenow is a PhD candidate in the Department of War Studies at King's College London. Her thesis engages with questions of political power and resistance based on the concepts and theories of Michel Foucault.

Anna Selmeczi is a PhD candidate in the Department of International Relations and European Studies at the Central European University. Her doctoral research focuses on the possible forms of resistance to (global) biopolitics. As part of this work, she recently conducted field research in the Abahlali base Mjondolo—a South African shack-dwellers movement.

Jorg Spieker is a PhD candidate in the Department of War Studies at King's College London. His dissertation examines the problem of war within the domain of liberal political theory.

Doug Stokes is a Senior Lecturer in International Relations at the University of Kent, Canterbury. His new book, *Global Energy Security and American Hegemony* (Johns Hopkins University Press, 2010) explore the role that US oil interventions are playing in maintaining US hegemony within the world system.

Jason R. Weidner is a PhD candidate in the Department of International Relations at Florida International University in Miami. He is currently researching neoliberal governmentality and subjectivity in the context of different projects for regional integration of the Americas.

Abstracts

Neoliberal Political Economy and the Subjectivity of Crisis: Why Governmentality is Not Hollow

NICHOLAS J. KIERSEY

This paper revisits Foucault's understanding of the importance of subjectivity for politics, focusing in particular on his claims concerning the sorts of demands placed on the subject by contemporary capitalism. Moves to extend the application of Foucault's analysis of liberal modernity to the realm of world politics have met with heavy criticism lately. According to David Chandler, for example, the idea of a "global governmentality" rests on the fundamentally unreliable premise that contemporary globalisation is driven by a kind of hyperbolic or imperial cosmopolitanism. Such arguments, he suggests, fail to recognise the progressive hollowing out or "attenuation" of the political that in fact makes liberalism impossible in the late modern era. In response to this argument, and recent similar arguments made by Marxists, this paper attends to what Foucault referred to as the "consciousness of crisis" that grounds the project of neoliberal governmentality. It is Foucault's contention that neoliberal capitalism has a consciousness of itself as a theory which seeks to incite entrepreneurialism to the point of crisis. In this sense, to speak of a crisis of political legitimacy is not to speak of a kind of mass passivity or "checking out", as Chandler might put it, but rather as an emergent form of behaviour that has been elicited or produced in a population. In order to escape Chandler's resentment of the failure of populations to live up to their responsibilities as political individuals, then, this paper draws on the theory of neoliberal governmentality. Understood principally as a summoning of entrepreneurial behaviours, the paper suggests that neoliberal governmentality affords us an opportunity for shifting the question of responsibility away from the sort of sovereign individualist platform elaborated by Chandler and towards a platform grounded more in the context of a dynamic and flexible global capitalism. To explore these issues, the paper contends, we should supplement Foucault's few rudimentary remarks on the subjectivity of neoliberal capitalism with the more fine-grained methods of "postliberal" economic analysis, such as that found in the works of Hardt and Negri.

Governmentality, Capitalism, and Subjectivity
JASON R. WEIDNER

This article takes as its starting point the important contribution that governmentality studies make to our understanding of the social and political conditions that shape contemporary world politics. However, it suggests that the critical potential of a governmentality approach can be more fully realised by dealing in a more substantive fashion with recent developments in capitalism and the latter's relationship with political subjectivity. The article introduces some elements of Italian autonomist Marxist thought and suggests that this intellectual tradition, together with Foucault's theorisation of neoliberal subjectivity in his recently translated 1979 lectures, can offer important insights that could strengthen governmentality accounts of contemporary social and political reality.

Governmentality of What? Populations, States and International Organisations
JONATHAN JOSEPH

As more work on governmentality appears in International Relations (IR), it is time to take stock and deal with a few questions. In social theory, the governmentality approach has mainly addressed "advanced liberal" societies and can be defined as having the health, wealth and well-being of populations as its target and governance from a distance through the "conduct of conduct" as its means of operating. There are two major problems in transferring governmentality to IR. First, not all societies can be described as "advanced liberal" ones. Second is the problem of whether there is such a thing as "global governmentality". This article argues that the lack of the necessary social conditions does indeed make it difficult to apply the technologies of governmentality to various parts of the world. However, the aim of international organisations might be less the regulation of populations as the application of governmentality to states.

Foucault's Concept of Power and the Global Discourse of Human Rights
IVAN MANOKHA

This article carries out a Foucauldian analysis of the global discourse of human rights. In the spirit of Foucault's genealogy it identifies a historical discontinuity in the development of human rights between the Cold War period, when human rights were a heavily contested concept, and the late modern world in which human rights are becoming a form of global standard whose validity is less and less questioned. Using Foucault's understanding of the way power is exercised within discursive structures and knowledge/power configurations, the article argues that in the late modern world human rights constitute a global norm with reference to which agents are evaluated and increasingly evaluate themselves. Power may be exercised over those forms of agency that do not conform to this norm by other agents, and at the same time, we witness more and more situations in which there is no such coercion and yet agents find it necessary to alter their behaviour and declare their adherence to human rights. The article also seeks to extend Foucault's framework in order to explain the emergence of the global human rights discourse. It thereby complements a Foucauldian analysis with a political economy approach which, it is argued, helps us enhance the

Foucauldian framework which suffers from the separation of discourses and politics from the economy.

Hobbes, War, Movement

LEONIE ANSEMS DE VRIES AND JORG SPIEKER

While informed by Foucault's understanding of power in terms of war and circulation, this article challenges Foucault's static reading of Hobbes. Contextualising Hobbes's political thought within the scientific ideas that he was inspired by, this article reveals that there is more to Hobbes than the static, depoliticising image of the contract. Hobbes's political thought is premised upon an ontology of movement; that is, his account of political order pivots on a double movement in which war constitutes the very possibility of social and political relations as well as of their continued reproduction via circulation. It is this conceptualisation of order that makes Hobbes's liberal political thought genealogically significant. And it is the model of a play of (re)productive movements—in certain respects close to Foucault's own conception of power—that can be used productively for thinking about governance and resistance today.

Taking Foucault beyond Foucault: Inter-state Governmentality in Early Modern Europe

HALVARD LEIRA

The analysis of governmentality has had a profound impact on the study of liberal, domestic societies over the last two decades, and the conceptual framework has been applied successfully to current global affairs. In this article one possible way of expanding the timeframe and the scope of governmentality studies is explored. Through an immanent critique of Foucault's own comments on the co-constitutive development of states and a state system in early modern Europe, it is argued that a governmentality perspective can in fact add to our understanding of inter-state relations in early modern Europe, and thus also to our understanding of our own time. Carrying out such analyses implies taking the Foucauldian framework beyond Foucault, as his own brief comments on inter-state relations fail to adhere to his own methodological precept of historicising seemingly evidentiary practices.

Decentring Global Power: The Merits of a Foucauldian Approach to International Relations

DOERTHE ROSENOW

In recent times, the value of a critical approach to the study of International Relations (IR) that makes use of the concepts and methods of Michel Foucault has (again) been put on trial. I will argue in this article that both Foucauldians and their critics often neglect Foucault's radical epistemology that always prioritises practices over political theory. The demand of such an approach is the relentless decentring and diversifying of totalising and unifying accounts of (global) power relations, resulting in a continuous challenge of the traditional meta-theories and concepts of any academic discipline—including IR. The present article will follow this approach and challenge, through the investigation of a particular case of what is commonly perceived as an exercise in "global governance", the idea that contemporary (global) power relations can be depicted solely through the lens of neoliberalism, sovereignty, or biopolitics. Instead, it will show that (global) power is located in a complex and flexible constellation of diverse and contradictory, mutually constituting and mutually destabilising strategies and tactics at particular sites.

"... we are being left to burn because we do not count": Biopolitics, Abandonment, and Resistance

ANNA SELMECZI

Starting from the puzzle posed by the ultimate aim of modern governmental rationality to nurture the population and its tendencies to exclude large parts of the same population from the spectrum of its care, this article argues that abandonment is always already inscribed into this rationality. In contradiction to Agamben, abandonment here is not attributed to the sovereign exception but is traced back to modern processes transforming the political—as problematised by Hannah Arendt and Michel Foucault. Complementing their observations with the empirical and the anti-political implications of "the count" based on Ian Hacking's and Jacques Rancie`re's thought, first a conceptual framework for understanding biopolitical abandonment is outlined, then the materialisation of abandonment is assessed. Arriving finally at the possibility of thinking resistance to the power that disallows life through conceiving of politics as disruption, the final section discusses the South African shack-dwellers' struggle that, on occasions, is able to disturb the dynamics of abandonment and so potentially furthers the conceptualisation of resistance to biopolitics.

Introduction[1]

NICHOLAS J. KIERSEY, DOUG STOKES AND
JASON R.WEIDNER

Foucault and IR

The pieces assembled in this volume are all motivated to some extent by a desire to return to Foucault and ask how we think might anew the conditions of possibility of political thought, practice, and living in the world today. Foucault is, of course, hardly a new intellectual interlocutor in International Relations (IR) theory. Two decades ago, his name had already been recruited by voices from the margins in a call for greater epistemological diversity and the recognition that many of the objects taken for granted within the discipline were themselves situated in, and conditioned by, a 'limit attitude.' It was then that we saw Foucault's name invoked, along with that of Derrida, in a direct attack both on the category of the sovereign subject of knowledge and the foreclosure within the discipline of any discussion concerning the constitution of that subject. The fundamental question raised in this attack thus concerned just how 'this sovereign voice of interpretation' was differentiated and legitimised such that it could be 'recognized, despite its historicity, as a pure and extrahistorical presence, a self-evident and identical voice of truth itself?'[2]

The works collected in this volume remain faithful to the spirit and substance of that original attack. The lesson remains as valuable today as it was twenty years ago. That is, that the nature of power is never quite obvious to us, neither in terms of its location nor how it seeks to involve us in its strategies of knowledge construction. Yet the works collected here seek to raise further questions, too. For the Foucault invited to the conversation then was something of an archaeologist.[3] Which is to say, he was deployed as an ally in what might be broadly understood as a 'poststructural' or 'postmodern' critique in order to challenge many of the basic assumptions that lie at the heart of mainstream realist and neorealist accounts of world politics. In the works of this volume, however, Foucault is invoked not so much by way of an attempt to re-fight previously fought disciplinary battles but, rather, to explore what *else* might be said about world politics. In this sense, these works have much in common with more recent attempts to develop insights from his work within the discipline. The works presented here cite, for example, what might be called the 'biosecurity' approach to world politics. That is, a critique, prompted by the emergence of transnational redundancies in the national security apparatuses of many states, and demonstrated most effectively in the context of the so-called 'Global War on Terror', which seeks to explore how contemporary strategies and tactics of social defence are operationalised by hegemonic understandings of 'what life is, and what it is for'.[4] Yet this is not all that has been said with Foucault in IR. As such, the essays presented here also take a cue from literature dealing more explicitly with 'global governmentality,' where attention is directed towards the myriad transnational projects that aim at ordering and governing human subjects and their relations.[5]

To be sure, biopolitics and governmentality have much in common, a fact which does not surprise us since both approaches draw much for their inspiration from aspects of Foucault's work. Yet the papers presented here also respect the differences in their domains of analysis, and in the conceptual frameworks they employ. While we do not wish to speak of different 'schools' of Foucauldian IR, we recognise that there exist serious differences in interpretation and application of Foucault's thought to the domain of world politics. Indeed, in analysing the discourses and practices of world politics from the perspectives of biosecurity and governmentality, Foucauldian IR raises a number of important implications not only for the way we study world politics but, as the following chapters attest, also for the kind of political possibilities we might imagine.

The power that Foucault labels *biopower* is distinguished from sovereign power by the fact that it 'exerts a positive influence on life' and that it 'endeavors to administer, optimize, and multiply' life by subjecting it to 'precise controls and comprehensive regulations.'[6] Where the traditional model of sovereign rule was grounded in 'the right to *take* life or *let* live,' the regime of biopower realigns this power with 'the right of the social body to ensure, maintain, or develop its life.'[7] Foucault delivers the full logic and force of this transformation in an oft-cited passage:

Wars are no longer waged in the name of a sovereign who must be defended; they are waged on behalf of the existence of everyone; entire populations are mobilized for the purpose of wholesale slaughter in the name of life necessity: massacres have become vital.[8]

Today this argument constitutes an important foundation of what we can call biopolitical IR, the 'biopolitics of security' school, or the biosecurity approach.[9] An important area of research within this approach focuses on the connection between politics, security, and war within an overall framework of the biopoliticisation of contemporary social and political life.[10] This line of investigation builds upon Foucault's analysis of biopolitics and security, particularly the argument that, as Campbell puts it, 'biopolitics arrives with the historical transformation in waging war from the defense of the sovereign to securing the existence of a population … [T]his historical shift means that decisions to fight are made in terms of collective survival, and killing is justified by the necessity of preserving life.'[11] This reading of the biopolitical and biopower has provided a powerful tool for analysing contemporary modes of war and practices of security, most notably those that have arisen within the context of War on Terror.[12]

While there are significant differences among the various scholars who study the connection between contemporary apparatuses of security and the biopolitical, a shared assumption is that we are currently witnessing not only a transformation in the nature of the biopolitical, but also an extension of the latter through the 'globalization' of apparatuses of security. However, while the related notions of biopolitics and biopower are at the heart of many critical accounts of contemporary world politics, there has been surprisingly little discussion about how these concepts can best be understood, what kinds of readings of contemporary social and political reality they lend themselves to, and what kind of politics they enable or foreclose. Neither has there been much discussion of how biopolitics should be distinguished from governmentality, or the relationship between the two; indeed at times the terms are used almost synonymously.[13] Part of this has to do with the ambiguous way which Foucault himself used the terms biopolitics and governmentality. However, as Merlingen has observed, there is also a disjuncture between governmentality studies and biopolitical approaches – not only in terms of their objects of study, but also in the scope of their theoretical claims and analysis.[14]

Governmentality studies were first developed by some of the scholars who had collaborated with Foucault and/or participated in his seminars and research projects of the late 1970s (e.g. Jacques Donzelot, Pascale Pasquino) and also by a group of mainly British social theorists for whom Foucault's notion of governmentality offered an attractive alternative to Marxist understandings of the social and political transformations associated with Thatcherism and the dismantling of the Welfare State.[15] Not satisfied with attempts to apprehend those transformations in terms of 'ideology' or 'hegemony' by Marxist and Gramscian approaches, a number of scholars drew on Foucault to analyze phenomena associated with the development of economic neoliberalism in terms of rationalities and technologies of government that were irreducible to static categories such as class or the state.[16] This approach produced two influential edited volumes – *The Foucault Effect* and *Foucault and Political Reason* – that helped give impetus to the development of an 'analytics of government' throughout a range of disciplines and fields of study.[17]

While some today speak of a 'global governmentality,' this is by no means an uncontroversial position.[18] Indeed, initially, the study of governmentality was oriented to problems and forms of government that were located *within* national societies. This is perhaps not surprising given that the principal political problem to which governmentality studies were formulated – the rise of neoliberalism – seemed to have developed within a limited number of national societies: initially Great Britain and the United States. Of course the social and political restructuring that took place first in Chile and later in a number of developing countries would seem to make this understanding of the development of neoliberalism problematic. On the one hand, the formulation of 'advanced liberalism' by Rose and others clearly, and problematically, presents the development of neoliberalism necessarily as a challenge to the Welfare State within 'post-industrial' societies, thus ignoring the relationship between neocolonialism, development, and governmentality.[19] Indeed, a serious shortcoming of much of the governmentality literature is its failure to consider the history of the development of governmental rule in colonial spaces.[20] On the other hand, the recent publications of Foucault's 1978 and 1979 lecture series should make it clear that, despite the commonly held view that Foucault focused solely on the domestic arts of governance,[21] he in fact understood the constitution of international relations as being inextricably related to the development of modern forms of governmentality.[22]

Foucault revisited

If extant investigations of world politics inspired by Foucault seem therefore to follow two discrete directions of inquiry, the biopolitical and the governmental, the chapters in this collection constitute an effort to pause and critically evaluate the contributions of this work to date, and to suggest some new avenues for consideration. While the chapters all have much in common, we suggest that for now it might be helpful to think of them as falling into four basic thematic groupings. The first two chapters, by Kiersey and Weidner respectively, examine the specific theme of capitalist subjectivity and how Foucault's 1978-1979 *Collège de France* lectures, recently published in English as *The Birth of Biopolitics*, mark an invitation for further reflection not only on how the emergence of governmentality was mediated by discourses associated with economic liberalism but also how more contemporary 'neoliberal' discourses appraise the potential of the marketplace as a 'technology of the self' for the production of healthy, entrepreneurial subjects.

This last is a point of some significance in debating the merits of Foucauldian IR today because many recent criticisms of the Foucauldian 'turn' in IR seem to have misread or strategically played down the fact that governmentality can refer to *any* programmatic attempt to govern subjects, whether by structuring the field of possibilities for subjects and/or influencing the development of the subject itself. Thus while more Marxist approaches decry 'global governmentality' for its reduction of everything to 'power' and, consequently, its elision both of the uneven nature of global development and the particularity of anti-capitalist struggle, Kiersey suggests that one of the major assets of the Foucauldian approach is the fact that it *does* start its analysis of governmental power, in the broadest sense of the term, with a regime of 'subjectification' wherein a 'free' self is called on to turn his- or herself into a subject. While this is not to suggest that all subjects are equally free in power, it is to suggest that the intentionality of power is immanent to a diversity of sites and scales of social interaction.

Confusion concerning the nature of governmentality and its relationship to the subject is also demonstrated in a recent critique by David Chandler, who has argued that scholarship on global biopolitics is guilty of a kind of wilful misrepresentation of the amount of subjective energy that might be available for any 'crusade' of liberal securitisation.[23] Chandler's argument in this regards looms large in many of the contributions to the collection, but one theme that emerges among the various responses is that Chandler engages in a somewhat caricatured reading of Foucauldian IR's understanding of liberalism. For as Foucault himself makes clear in *The Birth of Biopolitics*, liberalism is a mode of governmentality wherein government's power of intervention is understood as having natural limits which must be respected if the 'the internal rule of maximum economy' is to be achieved.[24] Thus, as Weidner suggests, it is imperative that we try to address more fully the relationship between subjectivity and broader structures and mechanisms of power, and analyse the various ways that capitalism shapes contemporary social and political reality. This question leads Weidner to engage – unusually for an IR theorist – with Italian autonomist Marxism, a body of thought that combines Foucault's notion of the biopolitical with a re-theorization of contemporary capitalism. Thus, suggests Weidner, concepts such as 'immaterial labour', 'real subsumption', the 'social factory', and 'general intellect' provide useful conceptual tools for considering how subjectivity has become crucial for capitalist production and valorisation, as well as for related structures of power and rule.

In the second thematic grouping, we find Jonathan Joseph and Ivan Manokha offering sympathetically critical readings of global governmentality scholarship. For Joseph, the dilemma is clear: governmentality was developed as a method of gaining insight into the power relations of advanced liberal societies, where power relations could function in conditions of relative freedom. While Joseph here limits his use of the term governmentality to the narrower, regime-specific connotation understood by Foucault, he nevertheless raises an important question: under what circumstances might we wish to deploy the concept of governmentality at a higher, transnational level? What particular practices might we identify to justify such a move? Noting that Foucault never relinquished the relevance of the more traditional 'sovereign' and 'disciplinary' modes of power in his triangulation of the different forces which can be brought into play in governmentality, Joseph claims we might delve a little deeper into the sort of role Foucault reserved for the operation of institutions of state power. The power of these institutions should be kept in mind especially if we are to contrast the relatively straightforward development of neoliberal rule through self-governance in advanced western societies with the rather different relations of power that enmesh much of the developing world. For Joseph then, neoliberal regimes of power can be identified operating in emerging markets insofar as the way their practices are organised, such as in the case of private security operating in Darfur or the practices of global development institutions, but full significance of these practices cannot be appreciated unless it is also understood that these regimes are imposed from a position of imperial exteriority.

Challenging Joseph to a certain extent, Manokha brings to bear a study of the progressive elaboration over the last twenty years of a multi-levelled regime of human rights protection. Importantly, he notes, Foucault's notion of power was intended from the outset as a challenge to mainstream conceptions of power in Political Science which tend to posit an actor or set of actors in possession of some capacity or other of power which they then use to affect their will over others. As a critique of such instrumentalist readings of power, Foucault suggested we might think of the actions by some agents over the will of others more as a form of mediation. Thus both the 'victims' and 'agents' of power both to some extent present themselves as evidence of the extent to which standards or norms of behaviour have been able to institute themselves. What is remarkable, for Manokha, is the extent to which evidence of the institution of human rights norms has soared in the aftermath of the Cold War and the relative disempowerment of alternative articulations of human rights norms occasioned by this event. Central to Manokha's account of this transformation is the substitutable nature of the actors which relay this discursive power, acting as they do from a variety of levels and in a variety of directions. Manokha's sorting of the evidence on this account leads him to conclude that global human rights today has an increasingly confessional nature and that this cannot be accounted for through a simple ontology of coercion or imposition.

In our third grouping of chapters we shift register away from our focus on governmentality theory and practice, and consider aspects of Foucault's own empirical work which seem to require further elaboration and development. Thus, for example, Ansems de Vries and Spieker revisit Foucault's genealogy of the emergence of liberalism, its concern with the management of aleatory nature of human political subjectivity. For Foucault, Hobbes looms large here – but, critically, it is a reading of Hobbes that appears to miss or gloss over major aspects of his more important contributions. The authors draw on the work of two of Hobbes key sources of inspiration, and on two contemporary readings of Hobbes by political theorists, all to make the point that Foucault exaggerates the 'static' dimension of Hobbesian ontology (that is, the ordered and rational nature of the founding contract) at the expense of the circulatory. To wit, suggest the authors, a more nuanced account of Hobbes's 'world' would admit both his understanding of human life as a field of asystematic and productive forces, and the role of *movement* in securing – though perhaps never fully – the potentiality of this field as ordered and rational. To be sure, it may be unusual to read

Hobbes as making the argument that civil society is very much a work in progress for Hobbes. And it is more unusual still to be lead to understand that a certain labour to produce a subject of movement or circulation is central to Hobbesian ontology. However, the intention here is not quite to rescue Hobbes but to suggest that his description of the foundational problem of liberalism reveals a deeper – and much more useful – thinking than Foucault's critique lets on.

If the Hobbes presented by Ansems de Vries and Spieker is a resource for rethinking the nature of anarchy, it seems that Foucault's own reading of inter-state relations may have been overly constrained. As Leira's paper argues, the governmentality approach has much to offer to the study of international relations but, unfortunately, Foucault's reliance of the concepts of *raison d'etat*, balance of power, and diplomatic relations suggests a certain inconsistency with his own theoretical principles. For how did these notions emerge? And how did the practices of diplomacy and war-making become taken for granted? Foucault says little. Thus, Leira observes, the Foucault of inter-state relations appears to undermine the Foucault of governmentality, offering a perspective on power that ignores the productive/positive dimension of power-knowledge that is so key governmentality. Indeed, one finds an understanding of the international that has much more to do with the 'repressive hypothesis' that Foucault so strongly wanted to distance himself from. By focusing on a selection of texts overlooked by Foucault in his own appraisal of the period of classic inter-state politics, from the seventeenth to the mid-nineteenth century, Leira points to the need to push 'beyond Foucault' and make the case for an actual governmentality of the inter-state period. Moreover, as Leira hints in summation, Foucault-inspired scholarship in IR (such as Der Derian, Bartelson, and Ashley) may already have set out some of the early steps such research might follow.

Moving to the final section of the collection, we return to governmentality but this time more to the specific practices and ambiguities of governmental power. For Rosenow, as with many of the contributions to this volume, the objective is to decentre what might otherwise be assumed as given. Thus she chides critics of Foucauldian IR, including Chandler, for the fact that they do not really 'get' Foucault. She notes for example that, in terms of epistemology, Chandler has far more in common with the sorts of naïve cosmopolitan accounts of globalization than the Foucauldian accounts he seeks to associate them with. Yet her piece is also distinguished by the fact that it proceeds from her defence of this decentred and decentring ontology in the abstract to an engagement with actual practices of international political economy, especially with regard to the World Trade Organization. Of course, critical scholarship from a diversity of methodological perspectives might be brought to bear upon the problematic nature of this organisation and its activities: Neo-Marxism, Agamben's critique of sovereignty, and even the more recent critique of biopolitical security in IR. For Rosenow, however, such approaches have a tendency to reduce questions of international governance to overarching schematics of analysis, thereby obfuscating the polyvalent practices that might be at work and, indeed, how these practices can be mutually implicated in the reproduction of governmentality.

Following a similarly inspired line of thinking, Selmeczi, our final contributor, poses the question as to whether contemporary governmentality must not also somehow presuppose the abandonment of large numbers of the population. How might we incorporate the 'extreme marginalization' so endemic to world politics today within a Foucauldian framework? Selmeczi is aware of other contributions which might make for more obvious sources of inspiration in analyzing this problem, such as Agamben's concept of *homo sacer*. Yet she meters her praise for this approach with a reticence to describe abandonment in terms of pure domination. The essential freedom of those who are superfluous to neoliberal governmentality is attested by a host of practices of resistance, such as lip-sewing and refugee poetry. These practices often emerge spontaneously in reaction to laws that would otherwise eliminate the very possibility of these subjects meeting their basic physical needs. For language to express the fierceness of the challenge these practices present to the naturalisation of politics, Selmeczi turns to Ranciere's aesthetic understanding of politics which shows how autonomous movements can disrupt the spatio-temporal orders defined by governmental rationalities.

Are we, in the end, simply trying to claim that there is such a thing as 'global governmentality'? Following Andrew Neal's conclusion in response to these chapters, such an object could hardly be said to exist outside of the set of practices that might constitute it. Governmentality, we recall once again, operates on two registers. On the one hand, it serves as a conceptual framework for analysing *dispositifs* of power regardless of the domain (madness, delinquency, sexuality, urban planning) or scale (the body, the community, the nation, the global) by focusing on the strategic and programmatic dimensions of power – the forms of knowledge, technologies, rationalities, and subjectivities that are assembled in order to shape and direct the relationships and activities of subjects. On the other hand, it deals with a specific form of political technology or art of government – liberalism or neoliberalism – which constitutes, broadly speaking, the overall horizon of political thought and action of 'our modernity', for which, Foucault argues, 'the problems of governmentality and the techniques of government have...become the only political stake and the only real space of political struggle and contestation.'[25] Operating in the interstices between this governmentality as a framework for the analysis of power then, and this governmentality as a horizon of admissible problems, these chapters seek not only to explore and challenge those objects of global political life which might otherwise be assumed to exist prior to their interpretation as *problems* of government, but also to see how the marginalised might discover possibilities, against all the odds, for their subversion.

Notes

1 The papers assembled in this edition are the result of a series of discussions, starting with a panel organized by David Chandler at the 33rd Annual Conference of the British International Studies Association, University of Exeter, December 13-17, 2008, and continuing with discussions across a range of panels at the 50th Annual Convention of the International Studies Association, New York, February 15-18, 2009. These chapters represent just a sample of viewpoints and arguments extended on those occasions. We join the authors in extending our genuine thanks to the organizers of those panels, and to Doug Stokes and the editors of Global Society for making this possible. Thanks also to the many reviewers for their detailed evaluations and encouragement. Finally we would like to acknowledge the diligent work of the journal's Editorial Assistant, Claire Tanner.

2 Richard Ashley, 'Living on Border Lines: Man, Poststructuralism, and War,' in *International/Intertextual Relations*, ed. James Der Derian and Michael J. Shapiro (1989), p. 262. See also Richard K. Ashley and R. B. J. Walker, 'Reading Dissidence/Writing the Discipline: Crisis and the Question of Sovereignty in International Studies,' *International Studies Quarterly* 34, no. 3 (1990), Jim George and David Campbell, 'Patterns of Dissent and the Celebration of Difference: Critical Social Theory and International Relations,' *International Studies Quarterly* 34, no. 3 (1990), Richard K. Ashley, 'Untying the Sovereign State: A Double Reading of the Anarchy Problematique,' *Millennium: Journal of International Studies* 17 (1988), ———,

'The Geopolitics of Geopolitical Space: Toward a Critical Social Theory of International Politics,' *Alternatives* 12, no. 4 (1987).

3 Michel Foucault, *The Order of Things* (London: Tavistock, 1970), ———, *The Archaeology of Knowledge*, trans. A.M. Sheridan Smith, 1st American ed. (New York: Pantheon Books, 1972).

4 See, principally, Michael Dillon and Andrew W. Neal, eds., Foucault on Politics, Security and War (New York, N.Y.: Palgrave Macmillan,2008), Julian Reid, The Biopolitics of the War on Terror: Life Struggles, Liberal Modernity and the Defence of Logistical Societies (Machester; New York: Manchester University Press, 2007), Michael Dillon, 'Governing Terror: The State of Emergency of Biopolitical Emergence ' International Political Sociology 1, no. 1 (2007).

5 See Nikolas S. Rose, Pat O'Malley, and Mariana Valverde, 'Governmentality,' *Annual Review of Law and Social Science* 2 (2006), Jacques Donzelot and Colin Gordon, 'Governing Liberal Societies: The Foucault Effect in the English-Speaking World,' *Foucault Studies*, no. 5 (2008). See also Graham Burchell, Colin Gordon, and Peter Miller, eds., *The Foucault Effect: Studies in Governmentality: With Two Lectures by and an Interview with Michel Foucault* (Chicago: University of Chicago Press,1991), Andrew Barry, Thomas Osborne, and Nikolas S. Rose, eds., *Foucault and Political Reason: Liberalism, Neo-Liberalism, and Rationalities of Government* (Chicago: University of Chicago Press,1996).

6 Michel Foucault, *The History of Sexuality: An Introduction* (New York: Vintage Books, 1990), 137.

7 Ibid., 139.

8 Ibid., 137.

9 See the 'Biopolitics of Security Network' website: http://www.keele.ac.uk/research/lpj/bos/index.htm>.

10 See, for example, Dillon and Neal, eds., *Foucault on Politics, Security and War.*

11 David Campbell, 'The Biopolitics of Security: Oil, Empire, and the Sports Utility Vehicle,' *American Quarterly* 57, no. 3 (2005): 949.

12 Dillon and Neal, eds., *Foucault on Politics, Security and War*, Reid, *Biopolitics of the War on Terror*, Dillon, 'Governing Terror: The State of Emergency of Biopolitical Emergence'.

13 See, for example, Majia Holmer Nadesan, *Governmentality, Biopower, and Everyday Life* (New York: Routledge, 2008), Michael Dillon, 'Governing through Contingency: The Security of Biopolitical Governance,' *Political Geography* 26 (2007).

14 Michael Merlingen, 'Monster Studies,' *International Political Sociology* 2, no. 3 (2008).

15 Rose, O'Malley, and Valverde, 'Governmentality.', Donzelot and Gordon, 'Governing Liberal Societies: The Foucault Effect in the English-Speaking World.'

16 Rose, O'Malley, and Valverde, 'Governmentality.'

17 Burchell, Gordon, and Miller, eds., *The Foucault Effect*, Barry, Osborne, and Rose, eds., *Foucault and Political Reason: Liberalism, Neo-Liberalism, and Rationalities of Government.*

18 Jan Selby, 'Engaging Foucault: Discourse, Liberal Governance and the Limits of Foucauldian Ir,' *International Relations* 21, no. 3 (2007), David Chandler, 'Critiquing Liberal Cosmopolitanism? The Limits of the Biopolitical Approach,' *International Political Sociology* 3, no. 1 (2009).

19 Arturo Escobar, *Encountering Development: The Making and Unmaking of the Third World* (Princeton, N.J.: Princeton University Press, 1995).

20 In fact, the notion of governmentality has been used productively to examine colonial and post-colonial social and political projects. See, for example, David Scott, 'Colonial Governmentality,' *Social Text*, no. 43 (1995), U. Kalpagam, 'Colonial Governmentality and the Public Sphere in India,' *The Journal of Historical Sociology* 14, no. 4 (2001), James Ferguson and Akhil Gupta, 'Spatializing States: Toward an Ethnography of Neoliberal Governmentality,' *American Ethnologist* 29, no. 4 (2005), Michael Watts, 'Resource Curse? Governmentality, Oil and Power in the Niger Delta, Nigeria,' *Geopolitics* 9, no. 1 (2004), Morgan Brigg, 'Empowering Ngos: The Microcredit Movement through Foucault's Notion of Dispositif,' *Alternatives: Global, Local, Political* 26, no. 3 (2001), ———, 'Post-Development, Foucault and the Colonisation Metaphor,' *Third World Quarterly* 23, no. 3 (2002).

21 See, for example Selby, 'Engaging Foucault.'

22 See the lectures of 22 and 29 March 1978 in Michel Foucault, *Security, Territory, Population: Lectures at the Collège De France, 1977-1978*, ed. Arnold I. Davidson, trans. Graham Burchell (New York: Palgrave Macmillan, 2007), 285-332.
23 Chandler, 'Critiquing Liberal Cosmopolitanism? The Limits of the Biopolitical Approach.'
24 Michel Foucault, *The Birth of Biopolitics; Lectures at the Collège De France, 1978-79* (New York: Palgrave Macmillan, 2008), p. 318.
25 Foucault, *Security, Territory, Population*, 109.

Neoliberal Political Economy and the Subjectivity of Crisis: Why Governmentality is Not Hollow

NICHOLAS J. KIERSEY*

This paper revisits Foucault's understanding of the importance of subjectivity for politics, focusing in particular on his claims concerning the sorts of demands placed on the subject by contemporary capitalism. Moves to extend the application of Foucault's analysis of liberal modernity to the realm of world politics have met with heavy criticism lately. According to David Chandler, for example, the idea of a "global governmentality" rests on the fundamentally unreliable premise that contemporary globalisation is driven by a kind of hyperbolic or imperial cosmopolitanism. Such arguments, he suggests, fail to recognise the progressive hollowing out or "attenuation" of the political that in fact makes liberalism impossible in the late modern era. In response to this argument, and recent similar arguments made by Marxists, this paper attends to what Foucault referred to as the "consciousness of crisis" that grounds the project of neoliberal governmentality. It is Foucault's contention that neoliberal capitalism has a consciousness of itself as a theory which seeks to incite entrepreneurialism to the point of crisis. In this sense, to speak of a crisis of political legitimacy is not to speak of a kind of mass passivity or "checking out", as Chandler might put it, but rather as an emergent form of behaviour that has been elicited or produced in a population. In order to escape Chandler's resentment of the failure of populations to live up to their responsibilities as political individuals, then, this paper draws on the theory of neoliberal governmentality. Understood principally as a summoning of entrepreneurial behaviours, the paper suggests that neoliberal governmentality affords us an opportunity for shifting the question of responsibility away from the sort of sovereign individualist platform elaborated by Chandler and towards a platform grounded more in the context of a dynamic and flexible global capitalism. To explore these issues, the paper contends, we should supplement Foucault's few rudimentary remarks on the subjectivity of neoliberal capitalism with the more fine-grained methods of "postliberal" economic analysis, such as that found in the works of Hardt and Negri.

Introduction

This paper revisits Foucault's understanding of the importance of subjectivity for politics, focusing in particular on his claims concerning the sorts of demands

*A draft of this paper was first presented at a panel entitled "The Uses of Global Governmentality", at the 33rd Annual Conference of the British International Studies Association, University of Exeter, 13–17 December 2008. The author wishes to thank David Chandler for organising that panel and inviting the interesting debate that took place that day. Thanks are also due to Doug Stokes, who first proposed this special issue of *Global Society*. Finally, the author would also like to acknowledge the valuable input of Jason Weidner, Ian Bruff, Stephen Legg, Mark Kelly, and two anonymous reviewers.

placed on the subject by contemporary capitalism. Moves to extend the application of Foucault's analysis of liberal modernity to the realm of "global" politics have met with heavy criticism lately. Some have adopted a more Marxist perspective to suggest that Foucault's methods are simply not cut out for the study of situations where power differentials are present, such as those common in world politics. Given that Foucault's project was focused on the paradox of power under conditions of freedom within domestic contexts, these critics aver, he would likely have had little to say about the structure of global capitalism and the imperialist subjection it thrusts upon the marginal nation-states of the world.[1] For some, then, the very idea of a "global governmentality" suggests a certain obfuscation of the patterns of quite *illiberal* capitalist exploitation that prevail in globalisation. However, if this critique seeks to insert a rhetoric of imperial power into our analysis of world politics, another critique breaks with this materialist orientation to challenge Foucauldian International Relations (IR) on the grounds that it invests far too much faith in the potential of post-territorial politics. As David Chandler argues, the premise upon which the global governmentality thesis rests, that contemporary globalisation is driven by a kind of hyperbolic or imperial cosmopolitanism, is fundamentally unreliable. For a progressive hollowing out or "attenuation" of the political effectively constrains the very possibility of any meaningful form of global liberalism today.[2]

Both of these critiques bring important challenges to the table for Foucauldian scholarship in IR. For example, in their demands for attention to the ongoing acts of ideological reproduction which help to hide the arbitrary, imperial nature of world politics, Marxist critics invite analysts of governmentality and biopolitics to account for the considerable role played by the imperatives of a global capitalist economy in shaping ideologies of security. Such work is important for, as I have argued elsewhere, economic considerations have received short shrift in much of the literature on the biopolitics of the War on Terror.[3] Similarly, Chandler challenges Foucauldian research to clarify just what is meant by the term "liberalism" and how exactly the different sorts of subjects recruited in the production of globalisation can be said to be "liberal". This said, however, both formulations seem to skip over central components of Foucault's ontology. For example, while it is clear that Marxist IR theorists broadly share Foucault's concern for the impossible demands that capitalism makes of the subject, they systematically overlook the reasons why Foucault refused to fetishise economic relations. Chandler, for his part, rebukes scholars of global biopolitics for exaggerating the subjective energies available to global liberalism. Yet he does so on the grounds that they gloss over the complete failure of contemporary "political elites" to create any coherent programme of globally shared values.[4] This claim is remarkable given that it belies such a decidedly liberal aspiration for sovereign

1. See, for example, Jan Selby, "Engaging Foucault: Discourse, Liberal Governance and the Limits of Foucauldian IR", *International Relations*, Vol. 21, No. 3 (2007); Jonathan Joseph, "The Limits of Governmentality", *European Journal of International Relations*, Vol. 16, No. 2 (2010).

2. David Chandler, "Critiquing Liberal Cosmopolitanism? The Limits of the Biopolitical Approach", *International Political Sociology*, Vol. 3, No. 1 (2009).

3. Nicholas J. Kiersey, "Scale, Security, and Political Economy: Debating the Biopolitics of the Global War on Terror", *New Political Science*, Vol. 31, No. 1 (2009).

4. Chandler, *op. cit.*, p. 67.

individual responsibility, an ontological category which Foucault spent a good deal of his career trying to refute.

In response to these arguments, this paper attends to what Foucault referred to as the "consciousness of crisis" that grounds the project of neoliberal global governmentality. It is Foucault's contention that neoliberal capitalism has a consciousness of itself as a theory which seeks to incite entrepreneurialism to the point of crisis. In this sense, to speak of a crisis of political legitimacy is not to speak of a kind of mass passivity or "checking out", as Chandler might put it, but rather as an emergent form of behaviour that has been elicited or produced in a population. The article thus draws on recently published transcripts from Foucault's lectures at the Collège de France in the late 1970s which show quite conclusively that Foucault was curious about the role of capitalist norms and values in the constitution of contemporary liberalism. Indeed, more importantly for this article, we can see in Foucault's short discussion of the American strain of neoliberalism, which is arguably the more ascendant version in the current context, a concern with the manner in which contemporary capitalism relies on the market as a potential vector for the solicitation of specific norms of individual responsibility.

To be sure, not all theories of global governmentality and biopolitics address the role of economic ideology in contemporary globalisation.[5] To some extent, this can be attributed to the fact that Foucault's only sustained treatment of the issue remained unpublished until recently. Nevertheless, the new lectures do present us with an interesting opportunity to revise our understanding of the sources of neoliberal governmentality's globalising impulse. As we shall see, Foucault's own discussion of neoliberalism suggests that the security sought by biopolitics is mediated by a fundamentally economistic horizon of thought. Otherwise expressed, it is less a discourse of cosmopolitan rights and more a discourse of economic utilitarianism that determines what may be said to constitute a secure life. Foucault was interested in the desire of neoliberalism to provide social security by eliciting an economic or entrepreneurial responsibility from its subject. Governmentality desires both to rule a human whose life it understands to be a fundamentally transactional phenomenon and to instil in that human a certain sense of what counts as responsible behaviour in the marketplace of his life.

Importantly, such a reading of contemporary liberalism stands in marked contrast with Chandler's assessment of the study of global biopolitics as simply a naïve engagement with cosmopolitan theory, and places Chandler's resentment of the failure of global populations to live up to their political responsibilities as individuals in a more nuanced context. Indeed, it shifts the very question of responsibility away from the sort of sovereign individualist platform elaborated by Chandler towards a platform grounded explicitly in the normative logic of global capitalism. At the same time, however, Foucault's reading of neoliberalism takes place more on the level of knowledge than ideology, displacing the need to

5. For example, considerations of economic ideology make little, if any, appearance in such commonly cited works as Julian Reid, *The Biopolitics of the War on Terror: Life Struggles, Liberal Modernity, and the Defence of Logistical Societies* (Manchester: Manchester University Press, 2006), Jenny Edkins and Véronique Pin-Fat, "Introduction: Life, Power, Resistance", in Jenny Edkins, Véronique Pin-Fat and Michael J. Shapiro (eds.), *Sovereign Lives: Power in Global Politics* (New York: Routledge, 2004); Wendy Larner and William Walters (eds.), *Global Governmentality: Governing International Spaces* (London: Routledge, 2004); Michael Dillon and Julian Reid, "Global Liberal Governance: Biopolitics, Security and War", *Millennium: Journal of International Studies*, Vol. 30, No. 1 (2001).

derive governmentality from capitalist relations of production per se. To explore these issues, I argue that we should supplement Foucault's few rudimentary remarks on the subjectivity of neoliberal capitalism with the more fine-grained methods of "postliberal" economic analysis, such as that found in the works of Hardt and Negri.

Debating Global Governmentality, Global Biopolitics

Literature in IR theory on the theme of global governmentality and the closely related notion of biopolitics is by now widely acknowledged.[6] Much of this literature takes its cue from Foucault's more explicitly political texts, such as his classic *History of Sexuality, Volume 1*, and, more recently, the 1977/78 Collège de France lectures, published in English as *Security, Territory, Population*.[7] To summarise these texts very briefly, in these various sources we find Foucault distancing himself from the Marxist understanding of the state as somehow being the servant of dominant class power, subordinating this question as merely a facet of a broader politics of subjectification in which the subject of government is "free" but implicated in his or her own "subjectivation", or self-government. Thus the subject is constituted or "led" to regulate its own "modes of action", both in the sense of how it seeks to govern its own margin of activity and seeks to influence the possible range of actions of others.[8] As part of this discussion, one finds Foucault making an argument that political life in liberalism is distinguished from earlier forms by a unique commitment to *deterritorialisation*. Thus, if governmentality refers to the "conduct of conduct", it achieves a certain hegemony in the 17th century in the desire to produce security by eliciting the sorts of behaviours that are deemed becoming of the idealised citizen of a territorial state. However, while this concern with territorial security never quite fades, in the 19th century it is subordinated more or less to a logic of government focused on the statist goal of, simply, the "administration of things" or the "right disposition of things".[9]

One key point to take from this focus on disposing things correctly is that modern government is far more interested in divining tactics and strategies for the maintenance of a certain order than in specifying the particular ends of that order. As Foucault notes, the stakes of governmentality revolve around "problems" and "techniques" of government—indeed, these are the "only real space of political struggle and contestation".[10] In IR theory, however, much of the work inspired by Foucault has focused on the ancillary concept of biopolitics,

6. See, for example, Iver B. Neumann and Ole Jacob Sending, "'The International' as Governmentality", *Millennium: Journal of International Studies*, Vol. 35, No. 3 (2007); Timothy W. Luke, "Governmentality and Contragovernmentality: Rethinking Sovereignty and Territoriality after the Cold War", *Political Geography*, Vol. 15, No. 6/7 (1996); Larner and Walters (eds.), *op. cit.*

7. Michel Foucault, *The History of Sexuality: An Introduction, Volume 1* (London: Vintage Books, 1990); *idem, Security, Territory, Population: Lectures at the Collège de France, 1977–1978* (New York: Palgrave Macmillan, 2007).

8. Michel Foucault, "The Subject and Power", in James D. Faubion (ed.), *Power* (New York: The New Press, 2000), p. 343. For discussion of the distinction between subjectification (also known, more simply, as subjection) and the more active, self-governing *subjectivation* see Mark Kelly, *The Political Philosophy of Michel Foucault* (New York: Routledge, 2009), pp. 87–89.

9. Michel Foucault, "Governmentality", in Faubion (ed.), *op. cit.*, p. 208.

10. Foucault, *Security, Territory, Population, op. cit.*, p. 109.

which many read as referring to the emergence of a specifically modern reflexivity of governmentality in relation to its own need for order and stability. Foucault himself sets the agenda for this argument by suggesting that modern liberal societies seek security not in the service of the state or any sovereign power or master but, rather, in the name of "the biological existence of the population" itself.[11]

This theme of biopolitics has played large in Foucauldian IR, especially in the analysis of globalisation and the transnationalisation of security imperatives under conditions of relative interdependency. Julian Reid, for example, has provided an in-depth study of the War on Terror on these grounds, citing emerging redundancies between the "homeland security" strategies of various nation-states as indicative of a kind of novel, hyperbolic transnationalisation of the logic of biopolitics. Here, Reid shows that the defence of society in biopolitical security is legitimated by notions of what life is, and what it is for.[12] Foucault, Reid suggests, believes that modern liberalism reads human life as a somewhat unpredictable quantity, which must be managed if it is to be secure. As such, in the management of the aleatory, the rationality of modern government takes its cue from the science of military strategy.

As Reid suggests, Foucault took the entry of formal military strategy into the human sciences as a significant development. This is clear, for example, in his regard for Clausewitz's teachings on logistics, or the manipulation and arrangement of forces on the battlefield. These principles bear a marked resemblance to the sorts of strategies for the arrangement of populations which soon thereafter came to be promoted by state governments. According to Reid's Foucault, then, the advent of the science of strategy also marks the entry of that science into modern power/knowledge relations as a "model discourse", or "strategic principle from which all other areas of discourse take their cue".[13] The argument here, then, is that "War on Terror" governmentality presupposes a kind of generalised tactics of logisticisation and that the ideal of this logistical life also serves as a master imperative in terms of the self-understanding that liberalism articulates in legitimising its own defence. Moreover, liberalism is non-discriminating in terms of just who may be said to be a part of this life.

One can see a similar understanding of the hyperbolic nature of contemporary liberalism in a range of sources. Agamben, for example, suggests that the War on Terror marks a transformation of "normal politics" today into a permanent and universal state of exception.[14] Similarly, Hardt and Negri speak of a permanent state of exception, describing this as an episode in what they call modernity's "civil war". Global biopolitics, they suggest, is beholden to an "imperial humanitarian" ethic which frames its wars as necessary in "the interests of humanity as a whole".[15] Dean, too, has noted how the pervasiveness of ontologically non-specific enemies has ruptured the very logic upon which the European Law of Nations was founded, replacing it with a world order based on the logic of "international

11. Foucault, *History of Sexuality, Volume 1, op. cit.*, p. 137.

12. Reid, *op. cit.*

13. Julian Reid, "Foucault on Clausewitz: Conceptualizing the Relationship between War and Power", *Alternatives*, Vol. 28 (2003), p. 4.

14. Giorgio Agamben, *State of Exception* (trans. Kevin Attell) (Chicago, IL: University of Chicago Press, 2005).

15. Michael Hardt and Antonio Negri, *Multitude* (New York: Penguin, 2004), p. 60.

civil war".[16] In each of these accounts, the domination of the securitising logic of biopolitics is said to encompass a globe-spanning regime of liberally imagined human life.

The Marxist Critique

This reading of the War on Terror as the expression of a kind of a global liberal logisticisation or a hyperbolic biopolitics is controversial, however. One very significant response to this argument has issued recently from Marxist IR critics. Jan Selby, for example, has argued that Foucault's core scholarly purpose was to disrupt liberal progressivism. As such, he was concerned with the "mechanisms and rationalities of governance and self-governance under conditions of formal freedom".[17] According to this view, then, authentically Foucauldian scholarship intentionally steers clear of contexts where such conditions of "formal freedom" are not present, such as that of the international. More specifically, it steers clear of the domain of international relations, which is defined fundamentally by the prerogatives of state powers. Selby assures us that this is not to advocate a return to the sorts of naïve unit-system analyses advocated by more realist approaches but, rather, to insist that we avoid at all costs any overstatement of the "unity, evenness and indivisibility" of the world order.[18] Echoing Stephen Gill, then,[19] Selby argues that Foucault can be used to analyse the "how" of power relations, but not the "why". More specifically, his point is that notions such as governmentality cannot be used to account for the "densely structured webs of social relations" that enshroud the world today, many of which were established under conditions of imperial rule and, as such, can hardly be said to count as relations of "formal freedom". Approaching these phenomena from the question of "why" they exist, it is clear that they are not born of global biopolitics. Rather, asserts Selby, they are attributable to the "ontological specificity" and "irreducibility" of the international. That is, they remind us that the international is not "coextensive with, or indistinguishable from, the domestic realm".[20]

While largely in agreement with Selby, Jonathan Joseph adds some useful nuances to the above line of argument. For Joseph, the key issue is rooted in this question of uneven development and the requirements of global capitalism to maintain a different set of governmental priorities in peripheral regions. Joseph accepts that the concept of governmentality is roughly commensurate with the trends towards deregulation and the retreat of state powers that are widely apparent in many social spheres today. Attendant to this, of course, has been the ascendance of neoliberal ideologies of economic government which have effected a remarkable shift in the locus of government away from government in the formal sense and "downwards" towards the population in general. Today, therefore, we find governmentality scholarship dealing with questions of

16. Mitchell Dean, "*Nomos* and the Politics of World Order", in Larner and Walters (eds.), *op. cit.*, p. 52.

17. Selby, *op. cit.*, p. 331.

18. *Ibid.*, p. 336.

19. Stephen Gill, "Globalization, Market Civilisation and Disciplinary Neoliberalism", *Millennium: Journal of International Studies*, Vol. 24, No. 3 (1995).

20. Selby, *op. cit.*, pp. 337–338.

how the subject is increasingly called on to govern itself, and to act responsibly. In neoliberalism, we learn, "people are told to take charge of their own well-being and take rational decisions to avoid social problems like unemployment and poverty".[21]

These developments clearly have a great deal of energy behind them in the Western world, and serve as an important stage in the development of governmentality. What Joseph protests, however, is that this version of governmentality should be used to describe situations where far more disciplinary modes of power are present. Echoing Selby above, he argues that the point is not to deny that neoliberal capitalism is distinguished by its reliance on an intense and non-corporeal form of power but to suggest that this form of power is not ubiquitous in globalisation. The all important question here, then, concerns the "conditions of possibility" that appropriately contextualise the emergence of such power. And here, just as with Selby, the answer lies in the logic of "geopolitics".[22]

Joseph's solution to the methodological dilemma of applying governmentality to globalisation is to suggest that Foucault's understanding of the state actually works in two directions. On the one hand, we have Foucault's classic understanding of state as a distraction from the essential business of studying power relations. That is, a reading of the state that emphasises its total contingency. On the other, we have a reading more consistent with the anti-essentialism of some strains of contemporary Marxism. By way of evidence for this argument, Joseph cites from Foucault's 1977/78 lectures, *Security, Territory, Population*:

> The state is what must exist at the end of the process of rationalization of the art of government. What the intervention of raison d'État must arrive at is the state's integrity, its completion, consolidation, and its reestablishment.[23]

Given that Foucault made this comment in the context of a discussion of the origins of the logic of *raison d'État*, defining this as an early albeit internally focused form of governmentality, the merits of this quote as evidence for Joseph's argument are shaky at best.[24] This point aside, however, Joseph's argument might simply be read as suggesting that an understanding of the state as an ontologically reliable locus of power allows for the construction of a supplementary theory of governmentality, which can then be set to work alongside that used to analyse the governance of subjectivation practised in advanced Western states. This *different* form of governmentality, then, is distinguished by the fact that it operates not so much through the self-rule of populations but, rather, through the enforcement of regimes over *states*. This move, suggests Joseph, makes for a stark contrast with the sorts of claims made in global

21. Joseph, *op. cit.*, p. 5.

22. *Ibid.*, p. 8.

23. Foucault, cited in *ibid.*, p. 9. See also Foucault, *Security, Territory, Population, op. cit.*, p. 287.

24. Indeed, it is worth recalling that Foucault had little or no time for the state as an analytic category, going so far as to claim "I must do without a theory of the state, as one can and must forgo an indigestible meal". See Michel Foucault, *The Birth of Biopolitics: Lectures at the Collège de France, 1978–79* (New York: Palgrave Macmillan, 2008), pp. 76–77. For further discussion, see Nicholas J. Kiersey, "World State or Global Governmentality? Constitutive Power and Resistance in a Post-imperial World", *Global Change, Peace & Security,* Vol. 20, No. 3 (2008), pp. 369–370.

biopolitics scholarship of the sort referred to above. For if a relatively free and self-governing civil society is one of the preconditions for the existence of governmentality then Foucauldian IR must surely acknowledge that even if such society could be proven to exist, it would be "patchy and uneven" at best.[25]

By way of an example to support his argument, Joseph turns to the structural adjustment policies of the IMF in Africa. Joseph concedes that the "drive" to elicit global entrepreneurialism may be "real". However, echoing previous contributions concerning the applicability of governmentality to contexts of classical imperialism from Hindess, and Larner and Walters, among others, Joseph sees the need for a methodological distinction between situations where governmentality emerges more through distributed modes of power, and situations where it is driven "from the outside".[26] This distinction is important because it suggests that the drive to governmentality can also encounter sites where conditions are such that its designs will *fail*. Thus we can identify the limits of the concept of governmentality itself and, as Joseph warns, these limits must surely haunt any effort to study "global governmentality" which does not give proper regard for the uneven distribution of liberal values in societies around the world. The analysis here must thus return to the question of the state which is, at the end of the day, the "main source of governmentality", and the "main promoter" of its various techniques.[27]

Joseph's argument is exemplary in so far as it both offers a cogent overview of governmentality literature and shows how the question of uneven development requires theoretical innovation in governmentality scholarship. And it is indeed clear from his argument that there is a need to identify the institutions of the state both as playing an important role in effecting transnational governmental consensus but also, potentially, as an obstacle or a point of resistance to the same. What might be better clarified, however, is how the state might be thought of as a "source" of governmentality. This question raises the stakes in Joseph's argument significantly, and not just from a Foucauldian perspective.[28] For, as Bruff notes, there is also a debate within Marxist IR on the extent to which social phenomena can be understood as autonomous or transcendent to society itself.[29] If an institution is defined in wholly autonomous terms, then the society to which it corresponds is necessarily reduced to a mere agent of its power. For Bruff, however, such reductionism is common not just in traditional Marxist approaches but also in much of the research inspired by Foucault. Claims for the legitimacy of such research are often staked on a purported "epistemological honesty" and the desire to avoid deterministic accounts of state power. Yet, says Bruff, these ontologies are actually quite totalising for, in Foucault's thinking, "power relations" are always in the final instance "the singular source of all human practice".[30] To wit, if power is "everywhere" then it is difficult

25. Joseph, *op. cit.*, p. 12.

26. *Ibid.*, pp. 14–15. See also Barry Hindess, "The Liberal Government of Unfreedom", *Alternatives*, Vol. 26 (2001); Wendy Larner and William Walters, "Introduction", in Larner and Walters (eds.), *op. cit.*

27. Joseph, *op. cit.*, p. 21.

28. See footnote 23, above.

29. Ian Bruff, "The Totalisation of Human Social Practice: Open Marxists and Capitalist Social Relations, Foucauldians and Power Relations", *British Journal of Politics and International Relations*, Vol. 11, No. 2 (2009).

30. *Ibid.*, p. 341.

to pose the question of unevenness, for the particular causes of any such unevenness, such as the obstruction potentially presented by legacy modalities of power, are all smoothed over as ubiquitous power relations.

In this sense, Bruff shares with Selby and Joseph a concern that Foucauldian research in IR asserts an uncompromising framework of the social and tends to elide the particular in human practice. This is not an argument that politics takes place in any number of constitutive contexts but, rather, it is to say that there is a "complexity and richness" in human practice which a generalised ontology of power cannot hope to capture. This complexity is grounded in the fact that, throughout history, human life has had a continuous and intimate relationship with the question of how its basic material needs should be met. As Bruff notes, echoing Gramsci, "human social practice is also conditioned by the values, norms and beliefs that comprise any conception of how to organize production", including, potentially, a range of "pre-capitalist" sources.[31] Our current capitalist regime is thus inevitably founded on a range of pre-existing social structures which must be somehow accommodated or negotiated. In this sense, while anchoring his account firmly in materialism, Bruff allows for the interplay of ideological traces from a range of historical modalities of production in the constitution of the "common sense" of contemporary power. Understood as orientations or dispositions towards production, then, this notion of common sense allows for a diversity of political processes while simultaneously anchoring them in an overall development of global capitalism.

As we shall see, however, Bruff's description of the "epistemological austerity" of the Foucauldian approach is problematic, based as it is on a highly selective reading of Foucault's arguments about power which completely ignores Foucault's focus on the immanence of power to processes of social constitution, and vice versa. Power, as such, may be "everywhere" but it is never in a position of exteriority, whether this exteriority be grounded in a transhistorical logic of production or otherwise. For our immediate purposes, however, Bruff's work is useful for its suggestion of the kind of metric of power that Marxist IR requires of Foucauldian analysis. While Selby and Joseph do not venture so explicitly into this argument, they do underscore the point that governmentality qua ontology of power is overly reductive. Otherwise expressed, they are anxious that governmentality cannot be scaled without reference to particular agencies of power that explain *why* world order develops unevenly. For his part, Bruff's work suggests that this anxiety also operates at a more meta-theoretical level. Foucauldian ontology is condemned to overlook the significance of basic material unevenness between societies and the various and particular "lived realities" of those societies because it cannot appraise the human being as anything other than a passive "vehicle for power relations".[32]

The Communitarian Critique

If the critique of Foucauldian IR presented in the last subsection was premised on the idea that Foucauldian ontology is ignorant of the importance of power differentials in the reproduction of globalisation, and necessarily so because of its

31. *Ibid.*, p. 346.
32. *Ibid.*, pp. 343–347.

tendency to divorce regimes of power/knowledge from the messy contexts of production and resistance which yield them, David Chandler seems a little more forgiving of the idea that we can confidently discuss globalisation as a globally emergent regime of norms and values. This said, like the Marxists, he is basically sceptical that this regime would best be referred to as a form of *liberalism*. For despite the proclivity of Foucauldian IR theorists to obsess over the imperialisms hidden in liberalism's global cosmopolitan aspirations, we should recognise that efforts to bring these aspirations to fruition have yet to achieve any real success. These efforts fail, suggests Chandler, neither because of the intractability of locally specific manifestations of governmentality nor because of the presence of counter-governmental resistances. Rather, as we shall see, they fail for a reason that is completely undetectable within the methodological parameters of global governmentality theory.

Before we detail Chandler's assessment of Foucauldian IR, it may be useful briefly to take stock of the reasoning behind his more general scepticism of theories of "global civil society".[33] For Chandler, the principle dilemma of the communicative approach to global civil society theorising is that it imagines a global "space" which encapsulates an emerging and transnationally articulated regime of "superterritorial" values. These values, such as respect for basic human rights and environmental security, are distinguished by the fact that they are not attached to the interests of traditional territorial state sovereignty, thereby transcending the limits of "territorial space". What has enabled this new space? Advocates of the space of global society often cite transformations in media technology. Inspired, we imagine, by the advent of global television networks, the Internet, and 24-hour news media coverage, these theorists explore the plausibility of a transformed global consciousness.

Chandler casts a wide net in identifying a spectrum of thinkers who buy into one or another version of this notion, grouping together cosmopolitan theorists such as Mary Kaldor with more radical critics like William Connolly and the "Multitude"-peddling Hardt and Negri. He cites the Critical Geopolitics scholar Gearóid Ó Tuathail, for example, who argues that "global space becomes political space. Being there live is everything. The local is instantly global, the distant immediately closes. Place-specific struggles become global televisual experiences."[34] Here, suggests Chandler, we find the idea of a revolution in communications technologies linked explicitly to what is described, significantly, as a "non-governmental space that comprises many hundreds of thousands of more-or-less self-directing ways of life".[35] To this general point, Chandler raises no significant objections. There is, to him, nothing new in the observation that social and economic interactions take places between diverse communities outside the realm of formal government. What does raise his ire, however, is that evangelists of this transnational consciousness suggest that we also find *political* relations being forged between these diverse communities.

While accepting that its various contributors do not all frame their argument the same way, Chandler nevertheless suggests that global civil society theorists tend to posit in their object of study an essential political commitment to the

33. David Chandler, "Deriving Norms from 'Global Space': The Limits of Communicative Approaches to Global Civil Society Theorizing", *Globalizations*, Vol. 4, No. 2 (2007).

34. Ó Tuathail, cited in *ibid.*, p. 285.

35. Keane, cited in *ibid.*, p. 287.

overcoming of the traditional territorial state model of citizenship in favour of a transnational or "non-exclusivist" model. In this way, global civil society theorists belie their own normative assumption that such a citizenship would necessarily constitute the emergence of a globally oriented ethics and the rejection of ideological dogmatism. The ability of such a citizenry to transcend "instrumentality and competing interests" would be facilitated, however, not so much by the fact that such citizens would all seek "good causes" but, as Chandler cites Habermas, by the fact that they would pursue a diversity of causes in an arena founded on "discursive designs" themselves modelled on an "egalitarian, open form of organization".[36] In this sense, the thesis of global civil society is an idealised interpretation, based on a "framework of communicative dialogue", of what "actually existing" global civil society must necessarily be in order for it to exist at all.

To a degree, then, Chandler's argument amounts to a concern that global civil society scholarship idealises its object of study, immunising it from the contaminating influence of any sort of power relations. Chandler's more emphatic critique, however, is that global civil society jumps far too quickly to the analysis of the moral reasoning of this object of study, before having made any real effort to provide evidence to support claims of its actual existence. In short, the thesis "relies on claims about the communicative interaction of global civic actors which have little connection to reality".[37] Indeed, its advocates appear almost to accept this, noting that there is little actual connectivity between the great majority of these actors. Thus scholars like William Connolly can simultaneously claim that the global civil space is possessed of a "thick political culture" while insisting that it is a completely de-centred phenomenon, composed of "diverse constituencies" acting in networks of "multidimensional connections".[38] For Chandler, this sort of argumentation is just too much, belying a preference for the "fictional" and for "academic verbiage" over the serious work of proving that a culture of global civil society exists at all.[39]

Following this line of critique, Chandler has more recently turned his attention to the "biopolitical approach" in IR.[40] Focused primarily on the sorts of imperial biopolitics arguments outlined above, he notes that these arguments tend to engage in a naïve reading of the sorts of cosmopolitan discourses just discussed. Taking seriously the claims of global civil society theorists concerning the enablement of "a new progressive liberal subject" which arises "from below", the global biopolitics critique is anxious that this argument runs the risk also of uncritically legitimating "new totalizing mechanisms of intervention and regulation from above".[41] Thus, despite their well-meaning intentions, the global civil society movement blurs the line between security and development, investing a new global "super-sovereign" imperative to which older territorial sovereignties may become subordinated.[42] Chandler cites scholarship from Reid, discussed

36. Habermas, cited in *ibid.*, p. 290.
37. *Ibid.*, p. 293.
38. *Ibid.*
39. *Ibid.*, pp. 293–294.
40. Chandler, "Critiquing Liberal Cosmopolitanism?", *op. cit.*
41. *Ibid.*, p. 56.
42. *Ibid.*, p. 57.

above, as an exemplar of this sort of discourse, but he also identifies Duffield, Jabri, and Douzinas.[43]

Chandler's problem with this line of argumentation is somewhat predictable given our discussion of his earlier work; just as the global cosmopolitan theorists derive their object of study more from their own normative aspirations than from the empirical evidence, so too the global biopolitics approach seems to identify the values of this same non-existent movement as its principal foe. Thus, as he notes, it is one thing to suggest that the "unitary assumptions" of the modern liberal democratic process have been undermined by social processes associated with globalisation but it is another entirely to claim that "new post-territorial forms of political community have been constructed in their stead".[44] Chandler concedes that such a breakdown in narrowly statist values may indeed be underway. He attributes this, however, not so much to an emergent or immanent multitudinal resistance enabled by a post-territorial sphere of communication. Rather, for Chandler, the crisis of political representation has been brought on by an altogether more pernicious development, the "attenuation of political contestation".[45] In this sense, Chandler does appear to accept one of the basic premises of global civil society and biopolitical research in IR. Namely, that globalisation has eroded state power, producing a democratic deficit. Yet he insists that this erosion has only been made possible because of a simultaneous "emptying out" of the political.

Significant numbers of liberal citizens are today refusing to participate in the traditional institutions of politics. This is happening, Chandler suggests, not because we are in a transitional moment in the development of a new global political horizon but, more simply, and more worryingly, because their "political elites" have been unable to "create projects of political meaning, able to cohere their societies or to offer a program of shared values".[46] Chandler's argument here, it should be noted, echoes the despondent tones of Robert Putnam's well-known critique of the decline of American political life, *Bowling Alone*.[47] Here, Putnam argues that contemporary capitalism has produced an intensification of individualism in American society, thereby eroding the willingness of the population to engage effectively in the practices of dutiful citizenship. To effect this argument, Putnam provides extensive data to suggest that Americans are spending less time engaged in a range of areas of social activity: eating fewer dinners together as families, joining fewer clubs and societies, signing fewer petitions, etc. These are the precise sorts of activities that once enthralled Alexis de Tocqueville and represented, to him, the source of common identity that gave a tremendous vibrancy to American democratic life. Today, however, for Putnam, "social capital" is declining in the United States.

43. See Vivienne Jabri, *War and the Transformation of Global Politics* (Basingstoke: Palgrave, 2007); Costas Douzinas, *Human Rights and Empire: The Political Philosophy of Cosmopolitanism* (London: Routledge Cavendish, 2007); Mark Duffield, *Development, Security and Unending War: Governing the World of Peoples* (Cambridge: Polity Press, 2007).

44. Chandler, "Critiquing Liberal Cosmopolitanism?", *op. cit.*, p. 58.

45. *Ibid.*, p. 56.

46. *Ibid.*, p. 67.

47. Robert D. Putnam, *Bowling Alone: The Collapse and Revival of American Community* (New York: Simon & Schuster, 2000).

Like Putnam, Chandler seems to feel the problem of contemporary politics is that it is afflicted with a certain "hollowing out" of the source of its traditional strength, a population of legitimising subjects bound together by "cohering values and sentiments".[48] To ignore this basic nature of "the problem" is to open up the possibility of its tragic reproduction at a more global level:

> Without the need to worry about the constitutive relationship between government (sovereign) and citizen, political community becomes entirely abstract. There is no longer any need to formulate or win adherence to a political program and to attempt to challenge or overcome sectional or parochial interests.[49]

Framed this way, it is very easy to sympathise with Chandler's reading of the biopolitical approach. Far from producing Reid's effervescent and imperialistic liberalism, then, ready to wage a universal war "over ways of life itself", we find only an epidemic of mass passivity or "bowling alone". In this sense, global governmentality theorisation is grounded in an empirically incorrect reading of the basic problem of contemporary liberal politics and, consequently, suffers from a hopelessly naïve outlook for the possibility of its resolution. Yet, as we shall see in the next section, whereas the Marxists appear to read Foucault's theory of the subject as a kind of uncompromising structuralism, Chandler offers a reading of global biopolitics which radically plays down just how "actively" populations participate in their political (and economic) worlds. For Foucault, political legitimacy is never solely a matter of the expression of an intentional allegiance. Rather, it must also be a matter of the constitutive power of government. To speak of a "crisis" of legitimacy, then, is not to speak of a kind of syndrome of mass passivity but, rather, a behaviour that has been elicited or produced in a population. This distinction in our analysis is crucial because it points to a fairly different understanding not only of the role of social power in effecting the crisis of the political but also, moreover, what is to be done about it. Whereas for Chandler the solution is to return somehow to traditional state power as the grounds for a legitimate and efficacious community, biopolitical theory finds the avenue of instrumental legitimacy rather narrow and even dangerous.

Governmentality Revisited

The critiques of Foucauldian IR outlined above do make for a welcome contribution to ongoing debate concerning the applicability of terms like governmentality and biopolitics to contemporary world politics. By challenging the applicability of these terms, they suggest that further attention be directed to the "actually existing" sites of globalisation to see not simply whether these terms have an empirical fit but also how they might, if necessary, be modified to satisfy the complaints of these critics. Yet while empirical analysis is surely necessary, it is also important to be clear about how Foucault used these terms and, more importantly, why. At its most basic, of course, Foucault's methods were motivated by a desire to be done with such notions as social structure and

48. Chandler, "Critiquing Liberal Cosmopolitanism?", *op. cit.*, p. 53.

49. *Ibid.*, p. 66.

atomistic agency. Borrowing from Nietzsche, he advocated a method of "effective history" which would render contingent "everything considered immortal in man".[50] Lavin clarifies the spirit of this method, suggesting that it is "predicated upon the idea that subjects are neither the unmediated expression of existing conditions nor atomistic autonomous beings independent from their conditions".[51] By way of a rebuttal to the critics of global governmentality, then, the following remarks constitute a reading of Foucault's understanding of genealogy with a view to showcasing two overlooked aspects of the global governmentality thesis. On the one hand, I hope to show that, despite the claims of Selby, Joseph and Bruff, there is much more to global governmentality than a totalising discourse of ubiquitous power that elides the particular in globalisation. Indeed, as I shall argue, Foucault is very clear in his descriptions of his model of power that *nothing* is ever finally determined by power. On the other hand, in response to Chandler's lament that the global governmentality thesis fails squarely on account of its poor empirical grip on the sorry state of liberal politics today, I suggest that neoliberal economic governmentality labours vigorously on the production of subjects of economic responsibility. To wit, while Chandler finds that the major problem of politics today is a refusal of political responsibility, this is not to suggest for a minute that governmentality has no purchase on this phenomenon.

If Foucault's work can be said to have a coherency or unity, he expressed this in his own words as an abiding concern with the problem of the "effects of power and the production of truth".[52] Less clear, however, are the precise parameters of his understanding of the nature of this relationship between power and truth. Foucault himself eschewed the ideal of a "general theory of what power is".[53] Moreover, there were breaks and disjunctures in the rhetoric Foucault used to describe power. One such break is said to have occurred in 1977, for example, when Foucault appears to shift away from the register of the model of war to describe power relations towards the register of government.[54] Nevertheless, certain strains of Foucault's understanding of power remain consistent throughout his works. Above all else, it seems, his theory of power was linked to a critique of the Western ontology of the subject which, he felt, immobilised the flexibility of political life. Traditional Western models of politics, he complained, tend to treat power as a "commodity" that would wield or be possessed by a "phenomenological" subject.[55] This understanding of the subject's relation to power introduces a problem, however, in so far as it portrays power as pure instrumentality. That is, it portrays power in terms of "a unique source of sovereignty", whether this be expressed in terms of "powerful individuals" or the social

50. Michel Foucault, "Nietzsche, Genealogy and History", in Paul Rabinow (ed.), *The Foucault Reader* (Harmondsworth: Penguin, 1984), p. 87.

51. Chad Lavin, *The Politics of Responsibility* (Urbana: University of Illinois Press, 2008), p. 22.

52. Foucault, cited in Kelly, *op. cit.*, p. 4; see also *idem*, "Truth and Power", in Faubion (ed.), *op. cit.*, p. 117.

53. Foucault, *Security, Territory, Population, op. cit.*, pp. 3–4; see also *idem*, "The Confession of the Flesh", in Colin Gordon (ed.), *Power/Knowledge: Selected Interviews and Other Writings, 1972/1977* (New York: Pantheon, 1980), p. 199.

54. See Arnold Davidson's introduction to Michel Foucault, *Society Must be Defended: Lectures at the Collège de France, 1975–76* (New York: Picador, 2003), p. xviii.

55. Foucault, "Truth and Power", *op. cit.*, p. 118; see also *idem*, *Society Must be Defended*, *op. cit.*, p. 13.

structures which might be derived as constraining them, thereby possibly ignoring other relevant facets of power.[56]

Crucially, then, Foucault was motivated by a concern that we might overlook the various ways in which power can create regimes of "subjectification", directing the will of the subject in different historical contexts. However, it should be noted, a further crucial aspect of Foucault's definition of power is that it pertains only to situations where a certain margin of *freedom* can be identified. Power is exercised, he notes, "only over free subjects". That is, for power to be exercised, the subject must be presented with a situation "in which several kinds of conduct, several ways of reacting and modes of behaviour are available".[57] Importantly, the threshold for freedom here is quite low. Thus, in the abstract at least, the subject of power is the antonym of the chained-up slave, for example, who has no possibility of mobility or escape. We see, then, that, for Foucault, power is not a unilateral "creation" of a subject but, rather, a solicitation of the self who then "turns him- or herself into a subject".[58]

This last point is especially important, considering the argument outlined by Bruff who, as noted above, claims that Foucault posits "power" as the "singular source of all human practice". To the contrary, Foucault clearly wants to reserve a sense in which there is something like a "self" that exists prior to the subject. Following Nietzsche, for example, Foucault explicitly identifies the human body as a material level or plane upon which power has to operate.[59] Indeed, for Kelly, this suggests that Foucault is faithful to an ontology of the subject where "it is the power of our sub-individual forces which are at stake: their prior relations to one another are acted upon from without".[60] Thus we might recognise in Foucault what Bruff overlooks. Namely, that Foucault intended this sort of theorisation of the relation between the subject and power as a corrective to accounts that end up posing power relations as completely autonomous from human agency. For Foucault, one's "free" choice is not always really power's choice. As Foucault states, power exists only where there is the potential for a resistance. Thus, in this sense, a certain freedom, capacity for resistance, or the potential for doing otherwise than power might intend, "comes first".[61]

A major point of consistency of Foucault's understanding of the subject thus coheres around this idea that subjectivity is elicited by power but is not exclusively dependent upon it. Yet, to the extent that the body is "free" it is also a terrain of power. Indeed, Foucault goes so far as to regard the body as something like a battlefield for clashing forces of truth which struggle over its domination. Foucault turns to the language of war to explain the dynamic relationship between truth and the body. In a piece that focuses heavily on the role of the body in the creation of subjectivity, Foucault argues that humanity "installs each of its violences in a system of rules and thus proceeds from domination to domination".[62] This argument, which Foucault terms "Nietzsche's hypothesis",

56. Foucault, *History of Sexuality, Volume 1*, *op. cit.*, pp. 92–93.

57. *Idem*, "The Subject and Power", *op. cit.*, p. 342.

58. *Ibid.*, p. 327.

59. Foucault, "Nietzsche, Genealogy and History", *op. cit.*, pp. 82–83.

60. Kelly, *op. cit.*, p. 101.

61. Michel Foucault, "Sex, Power, and the Politics of Identity", in Paul Rabinow and Robert Hurley (eds.), *Ethics: Subjectivity and Truth* (London: Allen Lane, 1997), p. 167.

62. Foucault, "Nietzsche, Genealogy and History", *op. cit.*, p. 85.

is in fact an inverted articulation of Clausewitz's famous aphorism that war is the continuation of politics by other means.[63] Politics is not literally the extension of war, however. The point, says Foucault, is that the genealogist should seek to explain power in terms of "the entry of forces".[64] Indeed, this echoes one of Clausewitz's major arguments, which was that a battle could never be resolved in a purely dialectical fashion. In a sense, for Clausewitz, there was no such thing as a total victory in a military conflict. The clash of forces necessarily contends with certain limits—political constraints, for example—which prevent a straightforward dialectic between the armies taking place. Certainly in the abstract one could imagine the idea of war as an "act of force" to which there is "no logical limit", but in the warp and woof of real-world events and politics, the ideal of such a "pure" war inevitably fragments.[65]

As no one side ever achieves pure victory in war, then, no one force or power ever achieves complete dominion over the body. War, then, is simply a model or abstract schematic for the operation of power. In this sense, as Kelly has usefully shown, Foucault's rhetorical commitment to the idea of war as a model of power was weak. The equation of power to war, or games,

> was never taken up by Foucault as a formal model at all, but rather only a metaphor Foucault uses tentatively, whereas government actually is an available model for the operation of power which Foucault takes up decisively.[66]

Thus, if the modelling of power on war is simply one formulation of Foucault's understanding of the abstract relationship between the subject and power, the more concrete formulation of this relationship is the theory of government to which Foucault would devote much attention in the last years of the 1970s. This is consistent, for example, with Colin Gordon's argument that Foucault construes the term "government" in at least two ways.[67] On the one hand, we have a "narrow sense", which is the most commonly used understanding of the term, referring to the historical emergence of a type of governmental rationality which displaces the model of sovereign power and law. On the other hand, however, we have a "wide" sense of the term which, in Foucault's words, refers to "the way in which one conducts the conduct of men" and constitutes "an analytical grid for these relations of power".[68] Here, then, *governmentality* refers to any relationship of power to conduct. Indeed, Foucault suggests that governmentality thus understood is valid for "the analysis of ways of conducting the conduct of mad people, patients, delinquents, and children". Moreover, as he goes on, it may also be useful for addressing phenomena such as "economic policy" or "the management of a whole social body".[69]

63. *Idem*, *Society Must be Defended*, *op. cit.*, pp. 15–16.

64. *Idem*, "Nietzsche, Genealogy and History", *op. cit.*, p. 86.

65. Carl Von Clausewitz, *On War* (New York: Dorsett Press, 1991), p. 84.

66. Kelly, *op. cit.*, p. 60.

67. Colin Gordon, "Governmental Rationality: An Introduction", in Graham Burchell, Colin Gordon and Peter Miller (eds.), *The Foucault Effect: Studies in Governmentality* (Chicago, IL: University of Chicago Press, 1991), p. 2.

68. Foucault, *Birth of Biopolitics*, *op. cit.*, p. 186.

69. *Ibid.*

The wide definition of governmentality thus suggests, simply, that the intentionality of power is immanent to diverse and reversible sites and scales of social interaction. Does this suggest that we are all "free" in those sites and scales? As we have discussed, this is clearly implied to some extent in Foucault's early modelling of governmentality on war. Yet even here there is no suggestion that we are all equals in power's regime. As Foucault wrote in 1977, "in so far as power relations are an unequal and relatively stable relation of forces, it's clear that this implies an above and a below, a difference in potentials".[70] Returning to Selby and Joseph's lament about the applicability of governmentality to global relations, then, we see that their argument is only partially accurate. While governmentality as a theory of liberal rationality *may* be limited to Western societies, as they argue, governmentality as a theory of power surely is not. To wit, Foucault argues that, in terms of scale of application, governmentality is "not confined by definition to a precise domain determined by a sector of the scale, but should be considered simply as a point of view, a method of decipherment which may be valid for the whole scale, whatever its size".[71]

Certainly there is a temptation, as Legg has noted in an insightful critique of Sassen,[72] among others, for one to reinscribe a certain structuralism in applying ontologies inspired by Foucault to globalisation. Yet, as Legg suggests, Foucault was not a particularly structuralist thinker. Foucault used terms like "assemblage" and, more particularly "dispotif", to evoke specific deterritorialisations and reterritorialisations of power:

> If one argues that territorialisation goes beyond simply establishing borders to dictating legitimate sovereignty within scales attributed to those borders, we can also consider assemblages to simultaneously be engaged in processes of re-, and de-, scaling of space through complex geographical governmentalities.[73]

These assemblages are dynamic, shifting and can overlap, allowing us to consider, for example, the governmentality of phenomena such as imperialism, or internationalism, or trafficking. But, more importantly, in distinction to Sassen's deliberately neutered version of the term, they are assemblages of *power*.[74] Taking power seriously, then, Legg summarises global governmentality as the study of "historically specific experiences, effects and affects of scale and the different forces of capital through which they are produced".[75]

Interestingly, by invoking the term "capital" here, Legg suggests that we might press a little further with the idea that governmentality in the narrow sense is also

70. Foucault, "The Confession of the Flesh", *op. cit.*, pp. 200–201.

71. *Idem*, *Birth of Biopolitics*, *op. cit.*, p. 186.

72. Saskia Sassen, *Territory, Authority, Rights: From Medieval to Global Assemblages* (Princeton, NJ: Princeton University Press, 2006).

73. Stephen Legg, "Of Scales, Networks and Assemblages: The League of Nations Apparatus and the Scalar Sovereignty of the Government of India", *Transactions of the Institute of British Geographers*, Vol. 34, No. 2 (2008), p. 239.

74. Sassen offers the perplexing disclaimer that she wants to use the term "assemblage" only in the "dictionary" sense of the term. See Sassen, *op. cit.*, p. 5.

75. Legg, *op. cit.*, p. 239.

somehow "global". This is particularly apposite in relation to Chandler's critique of the contemporary state of liberalism. To be sure, Chandler identifies a definite point of vulnerability in the global governmentality thesis vis-à-vis governmental rationality per se. To cite Neumann and Sending, for example:

> liberalism is a particular logic of governing—a form of power that is characteristic of modern society, which operates indirectly by shaping and fostering autonomous and responsible individuals.[76]

Yet the important question here is what possible basis there can be for the assertion of an autonomous and responsible individual of liberalism when, as Chandler suggests, the evidence seems to point, if anything, to liberalism's "hollowing out". For a possible answer to this question we might well turn our attention back to the realm of biopolitics and, more explicitly, to Foucault's linkage of biopolitics to the economic anthropology of capitalism. Foucault's limited commentary on the significance of economic rationality has not received very much attention in IR. Some, like Jessop, have noted, however, that Foucault cultivated an active curiosity about how the discrete mechanisms and practices of modern social life became subject to a "strategic codification" and "consolidation" under the banner of a globalised political economy.[77] This in fact is clearly Foucault's position in the 1978/79 lectures wherein he discusses the eventual displacement of the logic of *raison d'État*, with its unlimited breadth of internal objectives vis-à-vis the population, by what he refers to as "a new type of global calculation".[78] That is, a certain displacement of the logic of the state augured by the advent of political economy, or the discovery of the market as a site of "veridiction" of the nature of man.[79]

While much might be gleaned from these remarks on the globalising tendencies of neoliberal economic discourse, what is perhaps more relevant in relation to Chandler's argument is the way in which Foucault also alludes in these comments to the emergence of an ideal subject of economic government. Liberal government, he advises us, is not a directly determined "consequence" of the emergence of the idea of the market.[80] Yet early liberalism is to a significant degree an effect of the notion of the market, in so far as it is through the concept of the market and its intrinsically economic subject, *Homo oeconomicus*, that a certain domain of activity is first identified wherein the state would seek to avoid intervention if its goal of achieving "the internal rule of maximum economy" is to be achieved. It is thought that the market has a capacity to regulate itself, in freedom, and in a manner far more complementary to man's nature as an economic creature. In this sense, the rationality of early liberal government is grounded in the idea that while the state must be economically powerful, its ability to access this power is necessarily conditional upon a certain respect for this capacity of the market. Thus Foucault argues that the concept of the market comes to serve as

76. Neumann and Sending, *op. cit.*, p. 694.

77. Bob Jessop, "From Micro-powers to Governmentality: Foucault's Work on Statehood, State Formation, Statecraft and State Power", *Political Geography*, Vol. 26 (2007).

78. Foucault, *Birth of Biopolitics*, *op. cit.*, pp. 56–58; *idem*, "Omnes et Singulatim", in Faubion (ed.), *op. cit.*, p. 316.

79. Foucault, *Birth of Biopolitics*, *op. cit.*, pp. 31–32.

80. *Ibid.*, p. 320.

a central "principle and method of the rationalization of the exercise of government".[81]

Yet if early Liberalism emerges in a recognition of the internal limits of formal government, this is not to say that what liberalism desires is an overall *diminishment* of government. To the contrary, liberal government is subsequently taken up as a question of individual responsibility. As Foucault notes, by the middle of the 19th century the question of governmental regulation had been reframed somewhat, from the pure laissez-faire of the original Classical Liberals to a more interventionist stance. Citing from the German "Ordoliberal" political economist Röpke, writing in 1950, for example, we learn that economic liberalism has become a question of "shifting the centre of gravity of governmental action downwards".[82] The Ordoliberals were broadly opposed to state interventionism and strongly in favour of letting the price mechanism direct the economy. That said, they rejected the laissez-faire naturalism of the Classical Liberals. To the Ordoliberals, man was not, therefore, as imagined in the 18th century—a *natural Homo oeconomicus* best left to his own devices—but a form of life that must be incited, through social policy, to recognise a certain responsibility to engage in risk taking and entrepreneurialism.

One might consider here, too, the subject posed in the more American brand of neoliberalism, associated with Milton Freedman and the Chicago School. Where the Ordoliberals posited a separation between the realms of social and economic activity, which could be bridged through the inculcation of an entrepreneurial spirit, the American neoliberals rejected this separation, narrating the social as an economic sphere in the first instance. From this point of view, every form of human activity becomes readable as an essentially economic pursuit. Or, that is, any activity carried out in the pursuit of some expected future income.[83] Central here is the theory of human capital, or the idea that *all* labour, including wage labour, can be understood as a voluntary investment or entrepreneurial activity carried out in the individual pursuit of surplus value. Yet if in the American approach the players in the market are all understood to have a certain tendency to self-govern, this is not to say government has nothing to do. Even here, government must attend to the "rules of the game".[84] The purpose of the "law" is to make sure the game is distributed throughout society and played, allowing everyone in the society to behave as rationally as possible. Government acts on the market, or "market milieu", to make this happen.[85] This is important for, as Foucault tells us, it is not just "any behaviour whatsoever" which counts as economic behaviour. Rather, the definition of *Homo oeconomicus* here is limited to aspects of human behaviour which might be subject to these environmental interventions. Or, that is, subject to government.

To be clear, then, Foucault is arguing that the responsibility of the neoliberal subject—whether in the German or the American formulation—is to bring itself to labour in a self-enterprising fashion. As he notes, however, *Homo oeconomicus* in neoliberalism is an ideal type which must be produced in actuality. On the

81. *Ibid.*, p. 318.
82. Röpke, cited in *ibid.*, p. 148.
83. *Ibid.*, p. 224.
84. *Ibid.*, p. 260.
85. *Ibid.*, p. 259.

one hand, he is someone who "accepts reality".[86] Yet, on the other, *Homo oeconomicus* designates "the abstract, ideal purely economic point that inhabits the dense, full, and complex reality of civil society".[87] How does actually existing capitalism work to produce this subject? How might this ideal type relate to the actual practices of capitalism? Foucault does not tell us much. Certainly we can see in Foucault an understanding of *why* "biopower" is important for capitalism:

> This bio-power was without question an indispensable element in the development of capitalism: the latter would not have been possible without the controlled insertion of bodies into the machinery of production and the adjustment of the phenomena of population to economic processes.[88]

Here, Foucault clearly indicates capitalism's desire to develop an ideal subject. Yet what is perhaps not clear here are the biopolitical techniques themselves.

As Jason Read has recently suggested, Foucault's *Birth of Biopolitics* only hints at neoliberalism's dominance as a mode of subjection, or a "technology of the self".[89] Even the well-known exponents of the more laissez-faire American anarcholiberalism, for example, insist that there must be a certain effort or labour on the part of policy makers to ensure that the market is instilling not only the habits of "practice" of the ideally calculative subject but also a certain sense of "responsibility" towards these ends. Consider this recent observation by Becker in relation to the recent US subprime mortgage market meltdown:

> An important foundation of the philosophy behind the arguments for private enterprise, free economies, and free societies more generally, is that these societies rely on and require individual decision-making and responsibility.[90]

If, as Foucault suggests, American neoliberalism has a tendency to think in terms of manipulation of environmental conditions (enticements, disincentives, etc.), one can clearly see in Becker an appreciation of the salutary effect such manipulations might have on the subject over time. The subject he desires does not exist naturally but, rather, is one who must repetitively be brought back to the "various mental and physical capacities" demanded of him by the market.[91]

Importantly, then, by moving the question of governmentality in this direction, Foucault is calling our attention to forces of subjectivation which, while rarely discussed in IR, are commonly addressed in what Lavin has termed "postliberal economics".[92] Asking the question "who responds to global capital?", Lavin

86. *Ibid.*, p. 269.

87. *Ibid.*, p. 296.

88. Foucault, *History of Sexuality, Volume 1*, p. 141.

89. Jason Read, "A Genealogy of Homo-Economicus: Neoliberalism and the Production of Subjectivity", *Foucault Studies*, Vol. 1, No. 6 (2009), p. 30.

90. Gary Becker and Richard Posner, "The Erosion of Individual Responsibility", *The Becker-Posner Blog*, June 29 (2008), available: <http://www.becker-posner-blog.com/archives/2008/03/the_erosion_of.html>.

91. *Ibid.*

92. Lavin, *op. cit.*, p. 61.

borrows from Hardt and Negri to suggest that "the dispersed mode of production corresponding to an increasingly deterritorialized network of finance capital and immaterial commodities informs a subjectivity that is dispersed and lacking in traditional sovereignty".[93] For Read, such thinking bespeaks a "turn to production" in our theorisation of capitalism. By this, Read intends to suggest a broader interpretation than Marx's strictly economistic production. This is, in fact, "an expansive ontology of production" which is both *biopolitical,* in so far as it looks at the historical transformation of capitalist production to encompass language, subjectivity, affects, and desire, all of which are imaginably vital to the sorts of discipline that Becker would see instilled in the subject, and *ontological,* in so far as the transformation in production has the potential in turn to transform not just things but relations and subjects.[94]

In terms of this biopolitical dimension, then, the important move made by Hardt and Negri is that they identify a "historical transformation" in the place of life in sustaining capitalism. They use such terms as the "social factory" to suggest the difficulty today in distinguishing between the sphere of production as classically understood, based on the simple buying and selling of labour and commodities, and the reproduction of society broadly speaking. Labour today is not limited to the factory or the specific place of work (the "job") but extends out into any area of society that can be recruited into the production of value. In short, it describes any form of activity that arranges social life. For Negri, then, "production and society have become one and the same thing".[95] Or, in a more recent framing,

> Life is no longer produced in the cycles of reproduction that are subordinated to the working day; on the contrary, life is what infuses and dominates all production. In fact, the value of labour and production is determined deep in the viscera of life.[96]

Thus "late" or "postmodern" capitalism, as it is sometimes referred, seems to have made all aspects of life fair game for commodification. As such, given that all life is now "in capitalism", so to speak, it is no longer reasonable to talk about merely passive or "docilized" bodies—we have to talk about continuously self-governing bodies. Thus the significance of grounding our account of contemporary power in the realm of "affect", or desire, is clear:

> The excess of value is determined today in the affects, in the bodies criss-crossed by knowledge, in the intelligence of the mind, and in the sheer power to act.[97]

Affect is thus the main stake of the game, in so far as the market relies on a metric of responsibility in judging the individual both on the extent to which he may

93. *Ibid.*, p. 73.

94. Read, *op. cit.*, pp. 25–26.

95. Antonio Negri and Félix Guattari, *Communists Like Us: New Spaces of Liberty, New Lines of Alliance* (New York: Semiotext(e), 1990), p. 119.

96. Michael Hardt and Antonio Negri, *Empire* (Cambridge, MA: Harvard University Press, 2000), p. 365.

97. *Ibid.*, pp. 365–366.

wish to become an entrepreneur and the extent to which he can successfully lead others to *desire* the consumption of certain goods and services. More critically, however, we see how the normalisation of this metric may serve to elide the inequality and exploitative relations of power which capitalism actually presents. To wit, one's failure to desire becomes the basis of the accusation of irresponsibility.

Against Chandler's dismissive reading of global governmentality, then, we see discussed in Foucault's writings on the ascent of neoliberal economic mentality a rationality of self-government in which it is very difficult to recognise anything like a hyperbolic cosmopolitanism. Rather, we see a concern with an emerging form of economic rationality, and a way of imagining human life, all as a "point of departure" for an understanding of "the general framework of biopolitics".[98] In this sense, it appears that Foucault intended his account of liberalism as an important nuance to the question of biopolitics, or the politics pertaining to the fact of the "biological existence of the population". To modify Reid's argument somewhat, if biopolitics concerns the legitimisation of politics through a definition of what human life is and what it is for, it seems for Foucault that liberalism's answer to these questions is rooted firmly in economic thinking.

Chandler, of course, might reject this reading of liberalism as having very little in common with what passes as biopolitical research in IR theory or the field of Security Studies, but it is much more difficult to engineer from this the broader claim that Foucault would have little or nothing to say about the "crisis" of contemporary liberalism. A principal foundation of Foucault's approach to power is that it only "holds good" to the extent that it "induces pleasure".[99] The contribution of *Birth of Biopolitics* is that it demonstrates how discourses of neoliberal economics imagine the market as a technology of power which does just that! In this sense, Foucault opens the door to the kind of "post-liberal" economic analysis advocated by Lavin, above. In drawing our attention to neoliberalism's generalised metric of entrepreneurial responsibility, Foucault brings into relief the possibility that the "post-political" crisis that so upsets critics such as Chandler and Putnam might plausibly be explained by the emergence of relations of biopolitical capitalism, and not merely norms of passive individualism and selfish consumerism. Certainly it may be fair to talk about a "crisis", as indeed Foucault does, but this is a crisis which governmentality already understands quite well. For the risk-seeking life of the entrepreneur generates costs which are inextricably linked to neoliberalism's "consciousness of crisis".[100] Such costs are something to be managed, to be sure, for governmentality understands all too well how, left unmanaged, the obsessions of entrepreneurial life, or the need to "live dangerously", will occasionally spill over into problems.[101] But to the extent that they are acknowledged, such problems have little to do with a passive subject. To the contrary, they are simply externalities which must, when necessary, be managed on the margins as costs of "manufacturing freedom".[102]

98. Foucault, *Birth of Biopolitics, op. cit.*, p. 22.
99. *Idem*, "Truth and Power", *op. cit.*, p. 120.
100. *Idem, Birth of Biopolitics, op. cit.*, p. 68.
101. *Ibid.*, p. 66.
102. *Ibid.*, p. 65.

Conclusion

In governmentality, the dualisms of more structuralist approaches tend to blur into each other; there is no clean distinction between knowledge and power, between politics and economics, the sovereign and non-sovereign, or the domestic and the international. The "governing" agents of power are caught up in the same relations of truth as the subjects. Rejecting the categories both of the self-contained liberal subject and the passive victim of repression, Foucault posed a subject that was at once both situated within power and responsible for its reproduction. Critics of global biopolitics seem to ignore this centrally important aspect of the spirit of Foucault's work.

Marxists, on the one hand, claim that Foucault was only interested in studying a context of domestic freedom and that therefore he cannot be "scaled" to address the vicissitudes of IR. In doing so, however, they miss a core political thrust of Foucault's method. That is, rather than posing a subject that was always and everywhere somehow an automaton of power, he understood that subjectification was a process of warlike complexity that would vary according to prevailing conditions. Chandler, on the other hand, posits a passive and politically irresponsible individual as a foil for the critique of global governmentality without seriously engaging with the ontological foundations of that critique. As such, Chandler misses a plausible alternative diagnosis of the modern malaise. Foucault only hints at this in *Birth of Biopolitics*, but he clearly wants to suggest that within neoliberal governmentality's rhetoric of laissez-faire we can see the legitimation of a series of market-based technologies for the production of a very active subject indeed. A subject, that is, whose first responsibility is to the reproduction of itself as the rational, capital-bearing subject of *Homo oeconomicus*. Otherwise expressed, a subject who accepts that the "rules of the game" oblige him to commit forcefully to the improvement of his own human capital and position in the market.

In sum, then, we have a theory of power which accepts the particularity of context and a theory of the subject whose active self-government guarantees the reproduction of power. These two key facets of the global governmentality thesis attest to the fact that the biopolitical approach does not hope for a delimited or universal truth about the globalisation of government. Global neoliberal governmentality obviously does not, and cannot, work on a truly global population. Yet, as Lemke observes:

> Governmentality construes neoliberalism not just as ideological rhetoric, as a political-economic reality, or as a practical antihumanism, but above all as a political project that endeavors to create a social reality that it suggests already exists.[103]

Importantly, then, governmentality is a method of inquiry that allows us to understand the narrow, strategic vision of a governmental rationality along with the wider set of practices and intimate mechanisms of power that conspire in effecting this strategy.

103. Thomas Lemke, "Foucault, Governmentality, and Critique", *Rethinking Marxism*, Vol. 14, No. 3 (2002), p. 60.

Like Neumann and Sending, then, we may acknowledge that the idea of a global governmentality "cannot be taken uncritically from Foucault's writings".[104] Yet while it might be argued, with Selby, that we ought not simply "port" Foucault over to the study of "the international", this is not to say that his model of discursive power has no relevance to the study of the actually existing scales and assemblages of power in today's world politics. Governmentality, suggest Neumann and Sending, augured a "new mode of objectification" which emerged as a necessary "politico-epistemic" response to the power of the discourse of political economy.[105] In the context of a thickening set of global relations, however, it may now be possible to speak of:

> a global system of indirect forms of power that operates to guide, shape and foster specific types of not only states, but also other polities, as well as individuals. It sets up standards of behaviour for individuals and models of institutions to be implemented and followed by all good members of the international community.[106]

Thus, notwithstanding the important and helpful critiques of global governmentality addressed in this paper, it remains difficult to read our time either in terms of an overarching struggle between imperial oppressor and its subject or in terms of a simple "hollowing" of the will of the responsible subject of politics. As the submissions by Rosenow and Manokha elsewhere in this issue attest, evidence of the embeddedness of governmentality in a thickening set of transnational social relations can be observed in a diversity of issue areas. But the power of *governmentality* is perhaps nowhere more explicitly visible than in contemporary capitalism's strategy of producing crisis subjectivity.

This is perhaps the most cautionary point for the Marxists and for Chandler. The emergence of a set of de facto obligations for populations to perform as *Homo oeconomicus* bespeaks the pre-eminence of a radically self-governing subjectivity in contemporary capitalism. Can we say this subject is a ubiquitous or globally scaled phenomenon? As Neumann and Sending remind us, conceptions such as "global" and "international" are useful only to the extent that we keep in mind their nature as virtual potentials; they are "ideal-typical" and "forged out of stuff which is particular to a specific time and a specific space".[107] However, by drawing our attention to the entrepreneurial genealogy of neoliberal government, Foucauldian IR suggests how institutions of power might both idealise a global population as *Homo oeconomicus* and develop strategies to recruit that population in its own self-governance. What are the technologies of the self that might labour today to bring the subject of this project into existence? And what obstacles might this project encounter? By asking these questions, the critique of capitalism as biopolitics aims to explore just how mutually interdependent world order and capitalist imagination are today.

104. Neumann and Sending, *op. cit.*, p. 698.
105. *Ibid.*, p. 693.
106. *Ibid.*, p. 699.
107. *Ibid.*, p. 678.

Governmentality, Capitalism, and Subjectivity

JASON R. WEIDNER*

This article takes as its starting point the important contribution that governmentality studies make to our understanding of the social and political conditions that shape contemporary world politics. However, it suggests that the critical potential of a governmentality approach can be more fully realised by dealing in a more substantive fashion with recent developments in capitalism and the latter's relationship with political subjectivity. The article introduces some elements of Italian autonomist Marxist thought and suggests that this intellectual tradition, together with Foucault's theorisation of neoliberal subjectivity in his recently translated 1979 lectures, can offer important insights that could strengthen governmentality accounts of contemporary social and political reality.

Introduction

The political thought of Michel Foucault has provided the inspiration for an increasingly large body of work in the study of world politics and in the discipline of International Relations (IR). This increased prominence of Foucauldian accounts of world politics has led to some stock-taking[1] and also critical evaluations.[2] One of the more powerful concepts Foucault developed is the notion of governmentality. In the past two decades, the concept of governmentality has provided an important analytical framework for scholars throughout the different disciplines and fields of study in the human and social sciences. However, with the proliferation of the concept there is also the danger that it will lose its critical force. Thus, one of the main premises of this paper is that in order to make good on their critical potential, governmentality studies need to provide a stronger account of the connections between the specific practices of governmentality that have

---

*The author is grateful to Nicholas Kiersey, Francois Debrix, and two anonymous reviewers for their helpful suggestions on earlier drafts of this article.

1. Michael Merlingen, "Foucault and World Politics: Promises and Challenges of Extending Governmentality Theory to the European and Beyond", *Millennium: Journal of International Studies*, Vol. 35, No. 1 (2006), p. 104; Michael Merlingen, "Monster Studies", *International Political Sociology*, Vol. 2, No. 3 (2008).

2. David Chandler, "Critiquing Liberal Cosmopolitanism? The Limits of the Biopolitical Approach", *International Political Sociology*, Vol. 3, No. 1 (2009); Jonathan Joseph, "The Limits of Governmentality: Social Theory and the International", *European Journal of International Relations*, Vol. 16, No. 2 (2010); Jan Selby, "Engaging Foucault: Discourse, Liberal Governance and the Limits of Foucauldian IR", *International Relations*, Vol. 21, No. 3 (2007).

been the object of study and the broader dynamics and relations of power that together shape our political present.

More specifically, this paper will address two areas where the critical potential of governmentality studies can be strengthened: the relationship between subjectivity and broader structures and mechanisms of power, and an analysis of the various ways that capitalism shapes contemporary social and political reality. In fact, the claim made here is that both issues are very much related, as subjectivity is increasingly both produced by and also productive for contemporary capitalism. In order to development this argument, the paper engages with Italian autonomist Marxist thought that combines Foucault's notion of the biopolitical with a re-theorisation of contemporary capitalism. While by no means a unified or homogeneous body of political thought, Italian autonomist Marxism nonetheless has produced a number of common themes and conceptual innovations which, it will be argued, can be productive for a rethinking of subjectivity and capitalism from a governmentality framework. Antonio Negri and Michael Hardt's employment of Marxist thought is by now well known. However, many of their insights regarding recent transformations in capitalism have been obscured by their argument regarding the constitution of a global Empire. The paper thus discusses the work of other important autonomist thinkers, particularly Maurizio Lazzarato and Paolo Virno and their development of notions such as "immaterial labour", "real subsumption", "social factory", and "general intellect". It will be argued that these provide useful conceptual tools for considering how subjectivity has become crucial for capitalist production and valorisation, as well as for related structures of power and rule.

At the same time, while a reflective engagement with Italian autonomist Marxist thought can help shed light on recent transformations in capitalism, there are also a number of tensions between the former and governmentality approaches to world politics in general—tensions which make the wholesale importation of autonomist thought into a governmentality framework a problematic undertaking. Thus, part of the challenge is to build on the important insights regarding the nature of contemporary capitalist subjectivity and forms of power, while at the same time maintaining the important distinction that Foucault makes between capitalist social relations and governmental rationalities and technologies that, while often intersecting with the logics and structure of capital, cannot be reduced to the latter.

In fact an important potential point of articulation with post-Marxist theorisations of capitalism and subjectivity is provided by Foucault in his 1979 lectures on neoliberal governmentality. Foucault's account of the changing nature of *Homo oeconomicus* within liberal thought parallels in important ways central elements of Italian post-Marxist theorisations of the ways that subjectivity is made productive of value and forms of rule within contemporary capitalism. Thus, after discussing Italian post-Marxist contributions to our understanding of contemporary capitalism and the production and productivity of subjectivity, the paper turns to Foucault's analysis of neoliberal subjectivity in order to argue for a rethinking of the relationship between subjectivity and contemporary forms of power. Specifically, it will be argued that Foucault's theorisation of the reconfiguration of subjectivity within neoliberal governmentality opens up a new direction for studies of neoliberal governmentality and subjectivity and also makes possible—indeed, necessary—critical reflection on the intersection of capitalism and neoliberal governance.

Governmentality and Subjectivity

The concept of governmentality[3] has provided an important conceptual framework for studying the mechanisms and processes by which different projects of *government*—understood in the broad sense of "the conduct of conduct"[4]—are assembled and put to work in order to govern different aspects of social reality. Thus, within IR and other disciplines and fields of study with an international or global focus, governmentality studies have detailed the different political rationalities and technologies of governmental practice (surveillance, risk management, network governance, privatisation, best practices, competitiveness indexing and country benchmarking, public opinion)[5] as well as the different issues, spaces and domains of social life (the European Union, the integrated region, globalisation, global civil society, development, security, the failed state, peacebuilding and peacekeeping, immigration, refugees, AIDS, humanitarian need, the environment, global health)[6] that have been constituted as areas for governmental calculation and intervention.

3. The key references for this analytical framework remain the two edited volumes on governmentality: Andrew Barry, Thomas Osborne and Nikolas S. Rose (eds.), *Foucault and Political Reason: Liberalism, Neo-liberalism, and Rationalities of Government* (Chicago: University of Chicago Press, 1996); Graham Burchell, Colin Gordon and Peter Miller (eds.), *The Foucault Effect: Studies in Governmentality: With Two Lectures by and an Interview with Michel Foucault* (Chicago: University of Chicago Press, 1991).

4. Michel Foucault, "Governmentality", in Graham Burchell, Colin Gordon and Peter Miller (eds.), *The Foucault Effect: Studies in Governmentality: With Two Lectures by and an Interview with Michel Foucault* (Chicago: University of Chicago Press, 1991).

5. Colleen Bell, "Surveillance Strategies and Populations at Risk: Biopolitical Governance in Canada's National Security Policy", *Security Dialogue*, Vol. 37, No. 2 (2006); Rosalyn Diprose *et al.*, "Governing the Future: The Paradigm of Prudence in Political Technologies of Risk Management", *Security Dialogue*, Vol. 39, Nos. 2–3 (2008); Tore Fougner, "Neoliberal Governance of States: The Role of Competitiveness Indexing and Country Benchmarking", *Millennium—Journal of International Studies*, Vol. 37, No. 2 (2008); Hans Krause Hansen and Dorte Salskov-Iversen, "Remodeling the Transnational Political Realm: Partnerships, Best-practice Schemes, and the Digitalization of Governance", *Alternatives: Global, Local, Political*, Vol. 30, No. 2 (2005); Suzan Ilcan and Lynne Phillips, "Governing through Global Networks: Knowledge Mobilities and Participatory Development", *Current Sociology*, Vol. 56, No. 5 (2008); Hans-Martin Jaeger, "'World Opinion' and the Founding of the UN: Governmentalizing International Politics", *European Journal of International Relations*, Vol. 14, No. 4 (2008); Anna Leander and Rens van Munster, "Private Security Contractors in the Debate about Darfur: Reflecting and Reinforcing Neo-liberal Governmentality", *International Relations*, Vol. 21, No. 2 (2007).

6. Didier Bigo, "Security and Immigration: Toward a Critique of the Governmentality of Unease", *Alternatives: Global, Local, Political*, Vol. 27, No. 1 (2002); Brendan Donegan, "Governmental Regionalism: Power/Knowledge and Neoliberal Regional Integration in Asia and Latin America", *Millennium: Journal of International Studies*, Vol. 35 (2006); Stefan Elbe, "Risking Lives: Aids, Security and Three Concepts of Risk", *Security Dialogue*, Vol. 39, Nos. 2–3 (2008); Julia Elyachar, *Markets of Dispossession: NGOs, Economic Development, and the State in Cairo* (Durham, NC: Duke University Press, 2005); John Heathershaw, "Unpacking the Liberal Peace: The Dividing and Merging of Peacebuilding Discourses", *Millennium: Journal of International Studies*, Vol. 36 (2008); Wendy Larner and William Walters, "Globalization as Governmentality", *Alternatives*, Vol. 29 (2004); Tania Murray Li, *The Will to Improve: Governmentality, Development, and the Practice of Politics* (Durham, NC: Duke University Press, 2007); Ronnie D. Lipschutz, "Power, Politics and Global Civil Society", *Millennium: Journal of International Studies*, Vol. 33, No. 3 (2005); Timothy W. Luke and Gearóid Ó Tuathail, "On Videocameralistics: The Geopolitics of Failed States, the CNN International and (UN) Governmentality", *Review of International*

In some respects, "global governmentality" approaches to world politics share some common features with liberal theories of "global governance": both seek to describe and account for sources of rule and order in a context of globalisation and the absence of a clear state-based international order. However, with their focus on the political rationalities and technologies of government, governmentality accounts of contemporary forms of power and rule differ significantly from liberal theories of global governance, which tend to see the latter as necessary responses to certain problems that are global in nature and require supranational mechanisms and institutions.[7] Similarly, the critical analyses of neoliberal forms of governmentality produced by Foucauldian scholars also have much in common with some historical materialist approaches to the contemporary world order that see the latter as being shaped to a large degree by "disciplinary" forms of neoliberalism.[8] However, while the latter tend to see neoliberalism in terms of an ideology that attempts to justify the restructuring of relationships between nation-states and transnational agencies along lines favoured by transnational capital, governmentality scholars tend to view neoliberalism as a heterogeneous set of governmental practices which formulate different responses to the problem of rule—"who can govern; who is to be governed; what is to be governed, and how"[9]—that are not ultimately functional to the interests of any social group or class.

Moreover, governmentality accounts of world politics differ from many other theoretical approaches by placing the relationship between subjectivity and forms of government at the centre of their analyses. Liberal accounts of global governance tend not to deal with the issue at all. Historical materialist accounts of world politics that are inspired by the writings of Antonio Gramsci tend either to neglect the relationship between subjectivity and political power or subsume the latter under the category of hegemony, wherein consent is secured for social and political arrangements that benefit "the interests of big corporate capital and dominant social forces in the G7, especially in the U.S."[10] For governmentality scholars, however, there is a complex relationship between the structures and relations of power and processes of *subjectivation*—or the production of subjectivity.[11]

Political Economy, Vol. 4, No. 4 (1997); Michael Merlingen and Rasa Ostrauskaite, *European Union Peacebuilding and Policing* (Abingdon and New York: Routledge, 2006); William Walters and Jens Henrik Haahr, *Governing Europe: Discourse, Governmentality and European Integration* (New York: Routledge, 2005); Laura Zanotti, "Imagining Democracy, Building Unsustainable Institutions: The UN Peacekeeping Operation in Haiti", *Security Dialogue*, Vol. 39, No. 5 (2008).

7. James N. Rosenau, "Governance in the Twenty-first Century", *Global Governance*, Vol. 1 (1995).

8. Stephen Gill, "Globalisation, Market Civilisation, and Disciplinary Neoliberalism", *Millennium: Journal of International Studies*, Vol. 24, No. 3 (1995).

9. William Walters and Jens Henrik Haahr, "Governmentality and Political Studies", *European Political Science*, Vol. 4, No. 3 (2005), p. 290.

10. Stephen Gill, "Constitutionalizing Inequality and the Clash of Globalizations", *International Studies Review*, Vol. 4, No. 2 (2002), p. 48.

11. Some scholars distinguish between "1) 'subjectification' (*assujettissement*) or the ways that *others* are governed and objectified into subjects through processes of power/knowledge (including but not limited to subjugation and subjection since a subject can have autonomy and power relations can be resisted and reversed), and 2) 'subjectivation' (*subjectivation*) or the ways that individuals govern and fashion *themselves* into subjects on the basis of what they take to be the truth". Trent H. Hamann, "Neoliberalism, Governmentality, and Ethics", *Foucault Studies*, No. 6 (2009) pp. 38–39,

Thus one of the major contributions of governmentality studies can be found in the many nuanced and fine-grained analyses of processes of subjectivation associated with neoliberal governmentalities, oftentimes employing thick ethnographic descriptions of the diverse sites and contexts within which neoliberal subjectivities are mobilised as part of specific governmental projects.[12] However, oftentimes these accounts of neoliberal subject formation are based on problematic understandings of subjectivity. To begin with, governmentality studies tend to examine discourses produced by particular authorities, promoting certain neoliberal aspects of subjectivity—responsibility, flexibility, rational calculation—without being able to account for the success or failure of that discourse to in fact produce the desired form of subjectivity. One is left, then, either to assume the efficacy of these discourses or to wonder exactly how they work. In other words, if governmentality approaches are based in part on the claim of a connection between neoliberal forms of governmentality and the production of neoliberal subjectivities, it would seem that they must produce some way of accounting for the mechanisms with which these subjectivities are produced.

Furthermore, despite what appears as a dynamic conceptualisation of subjectivity contained within governmentality accounts that look at the ways subjectivity is produced and modified, these accounts nevertheless tend to rely on a substantialist image of the subject,[13] so that neoliberal subjectivity is seen as a *thing* that can be produced (even if it is later altered). Thus governmentality studies tend to view neoliberal subjectivation in terms of the interiorisation of certain neoliberal values, attitudes, and beliefs. Of course, alternatives to this understanding of subjectivity have been presented by thinkers such as Judith Butler and others who stress the embodiment, performance, and enactment of subject positions in everyday situations.[14] Given that much of this literature already builds on some of Foucault's

footnote 4. In this paper I use the second term, subjectivation, to refer to the process of subject formation, whether by the self or by others.

12. Elyachar, *op. cit.*; Banu Gökarıksel and Katharyne Mitchell, "Veiling, Secularism and the Neoliberal Subject: National Narratives and Supranational Desires in Turkey and France", *Global Networks*, Vol. 5, No. 2 (2005); Banu Gökarıksel and Anna Secor, "New Transnational Geographies of Islamism, Capitalism and Subjectivity: The Veiling-fashion Industry in Turkey", *Area*, Vol. 41, No. 1 (2008); Wendy Larner, "Globalization, Governmentality and Expertise: Creating a Call Centre Labour Force", *Review of International Political Economy*, Vol. 9, No. 4 (2002); Li, *op. cit.*; Katharyne Mitchell, "Educating the National Citizen in Neoliberal Times: From the Multicultural Self to the Strategic Cosmopolitan", *Transactions of the Institute of British Geographers*, Vol. 28, No. 4 (2003); *idem*, "Neoliberal Governmentality in the European Union: Education, Training, and Technologies of Citizenship", *Environment and Planning D: Society and Space*, Vol. 24, No. 3 (2006); Mark Olssen, "Understanding the Mechanisms of Neoliberal Control: Lifelong Learning, Flexibility and Knowledge Capitalism", *International Journal of Lifelong Education*, Vol. 25, No. 3 (2006); Julia Paley, *Marketing Democracy: Power and Social Movements in Post-dictatorship Chile* (Berkeley: University of California Press, 2001); Katharine N. Rankin, "Governing Development: Neoliberalism, Microcredit, and Rational Economic Woman", *Economy and Society*, Vol. 30, No. 1 (2001); Daromir Rudnyckyj, "Spiritual Economies: Islam and Neoliberalism in Contemporary Indonesia", *Cultural Anthropology*, Vol. 24, No. 1 (2009).

13. Walter A. Davis, *Inwardness and Existence: Subjectivity in/and Hegel, Heidegger, Marx, and Freud* (Madison: University of Wisconsin Press, 1989), p. 20.

14. Karen Barad, "Posthumanist Performativity: Toward an Understanding of How Matter Comes to Matter", *Signs*, Vol. 28, No. 3 (2003); Judith Butler, *Bodies That Matter: On the Discursive Limits of "Sex"* (New York: Routledge, 1993); *idem*, *Excitable Speech: A Politics of the Performative* (New York: Routledge, 1997); *idem*, *The Psychic Life of Power: Theories in Subjection* (Stanford: Stanford University Press, 1997); John Cromby, "Theorizing Embodied Subjectivity", *International Journal of Critical Psychology*, Vol. 15 (2005).

insights regarding subject formation, it should not be difficult to incorporate embodied, performative approaches to subjectivity into governmentality analyses.

In fact, while this seems a necessary step towards strengthening governmentality accounts of subjectivity, it does not by itself address what I wish to argue is another shortcoming of governmentality studies—the elision of the relationship between capitalism and subjectivity, particularly as the former relates to structures and relations of power. This may seem to contradict the argument made above that one of the strengths of governmentality accounts of subjectivity is that they do not reduce the latter to the requirements of capital. However, it would seem that in their zeal to offer non-reductionist accounts of subjectivity that view neoliberalism as distinct and irreducible to contemporary forms of capitalism, governmentality studies have gone to the other extreme of ignoring the ways in which contemporary capitalism and neoliberal forms of government intersect and resonate with each other, as well as the ways in which the subjects of contemporary capital and neoliberal subjects often are subjected to similar mechanisms of control—indeed, the two are often indistinguishable. Here an engagement with Italian post-Marxism, particularly their theorisations of recent transformations in capitalist social relations and forms of power, can provide some critical tools for governmentality scholars to push their analyses in more productive directions.

Italian Post-Marxist Thought: Subjectivity and Capitalism

For the most part, scholars working from some kind of Foucauldian framework—whether from a biopolitical or governmentality perspective—have not made capitalism a central part of their analysis. What is more, their work has often reproduced the idea that there is a fundamental opposition between Foucault's work and Marxist thought. What is often ignored is that there already exists a group of thinkers who have drawn on Foucault's work in order to formulate theoretical and political responses to the current capitalist order. Beginning in the 1970s, Foucault's work inspired a number of critical Italian theorists to rethink the nature of capitalist development and capitalist social relations, leading some to speak of an "Italian Foucault".[15] The most prominent of these thinkers is Antonio Negri, whose work together with Michael Hardt[16] has generated a great deal of discussion in a number of fields, including IR.[17] Hardt and Negri's thesis regarding the emergence of a new form of global sovereignty—"Empire"—is by now quite well known and needs no further elaboration here.

15. Mark Coté, "The Italian Foucault: Subjectivity, Valorization, Autonomia", *Politics and Culture*, No. 3 (2003). There is in fact a massive literature on Italian social and political thought that engages with Foucault. Unfortunately, most of it remains untranslated.

16. Michael Hardt and Antonio Negri, *Empire* (Cambridge, MA: Harvard University Press, 2000); *idem, Multitude: War and Democracy in the Age of Empire* (New York: Penguin, 2004).

17. See, for example, the discussion in the journal *Millennium*: Tarak Barkawi and Mark Laffey, "Retrieving the Imperial: Empire and International Relations", *Millennium—Journal of International Studies*, Vol. 31, No. 1 (2002); Alex Callinicos, "The Actuality of Imperialism", *Millennium—Journal of International Studies*, Vol. 31, No. 2 (2002); Martin Shaw, "Post-imperial and Quasi-imperial: State and Empire in the Global Era", *Millennium—Journal of International Studies*, Vol. 31, No. 2 (2002); R.B.J. Walker, "On the Immanence/Imminence of Empire", *Millennium—Journal of International Studies*, Vol. 31, No. 2 (2002). See also the helpful critical discussion of *Empire* in Paul A. Passavant and Jodi Dean (eds.), *Empire's New Clothes: Reading Hardt and Negri* (New York: Routledge, 2004).

Less well known, however, is the large body of critical theoretical work that has provided innovative accounts of contemporary capitalism through a re-reading of the Foucauldian concepts of biopower and biopolitics. While a thorough review and discussion of this work is beyond the scope of this paper, I do wish to highlight a couple of elements of this area of scholarship in order to suggest how it can enrich already existing Foucauldian accounts of world politics—particularly by incorporating capitalism and its connection with subjectivity into the analytical framework.

The body of political thought I discuss here has been referred to variously as post-Marxist, workerist, post-workerist, and autonomist Marxism. The thinkers mentioned in this piece all have connections to the workerist political movement (*Potere Operaio*) that was active in the 1960s and early 1970s in Italy, and out of which was born the *autonomia* movement of Italian workers, students, and feminists.[18] Here I will use the terms "post-Marxist" and "autonomist Marxism" synonymously to describe critical thinkers who have been associated with these movements in Italy and have also sought to move beyond traditional Marxist political thought in trying to apprehend recent transformations in capitalism.

One of the foundational premises for autonomist Marxism is the inversion of the idea found in many Western accounts of capitalism that emphasise "only the dominant and inexorable logic of capital. Its accumulative logic, unfolding according to ineluctable (even if finally self-destructive) laws, figures as the unilateral force shaping the contemporary world".[19] Against this idea, autonomist Marxists have emphasised the creative power of labour and the reactionary nature of capital.

> Far from being a passive object of capitalist design, the worker is in fact the *active* subject of production, the wellspring of the skills, innovation, and cooperation on which capital depends. Capital attempts to incorporate labour as an object, a component in its cycle of value extraction, so much *labour power*. But this inclusion is always partial, never fully achieved. Labouring subjects resist capital's reduction. Labour is for capital always a problematic "other" that must constantly be controlled and subdued, and that, as persistently, circumvents or challenges this command.[20]

Related to this is the idea that the struggle by workers to escape the command of capital and achieve autonomy provokes capital to restructure the production process and the division of labour in order to reassert its command. This is captured by the notion of *class composition*, which "addresses both the 'technical'

18. Many of the issues related to workerist and autonomist thought are discussed in the special issue "Italian Post-workerist Thought", in *SubStance*, Vol. 36, No. 1 (2007). For a history of this political and intellectual movement, see Steve Wright, *Storming Heaven: Class Composition and Struggle in Italian Autonomist Marxism* (London: Pluto Press, 2002). Many of the political and theoretical debates that surrounded this movement are discussed in Sylvère Lotringer and Christian Marazzi (eds.), *Autonomia: Post-political Politics* (Cambridge, MA and London: Semiotext(e), 2007). See also Michael Hardt, "Introduction: Laboratory Italy", in Paolo Virno and Michael Hardt (eds.), *Radical Thought in Italy: A Potential Politics* (Minneapolis: University of Minnesota Press, 1996).

19. Nick Dyer-Witheford, *Cyber-Marx: Cycles and Circuits of Struggle in High-technology Capitalism* (Urbana: University of Illinois Press, 1999), p. 65.

20. *Ibid.*

organisation of the working class within the capitalist mode of production and the forms of 'political' subjectivity that emerge within the working class".[21] This reworking, or inversion, of the notion of the composition of class and its relation to the composition of capital has led to the argument that

> given a certain level of capitalist development, transformations in political composition precede and determine transformations in technical composition. This means that the labour struggles accompanying the emergence of a new political subject force entrepreneurs to implement processes of economic restructuring. In the workerist interpretation, such economic restructuring constitutes an attempt to contain the revolutionary transformations that capitalist development produces and depends on, even as these transformations continually call into question capitalism's material foundations, private ownership of the means of production, and entrepreneurial command over alienated labour.[22]

The importance of the notion of class composition came to be felt by the Italian workerist movement when the struggles that had been largely organised around the large industrial factories in the 1960s gave way to new strategies by capital such as the introduction of technological innovations and organisational restructuring in order to lessen capital's dependence on organised labour. At the same time, "student struggles and the emergence of both a highly combative feminist movement and new, strongly politicized youth cultures had shifted the terrain of struggle away from the factory".[23] Taken together, these two developments—the reorganisation of capitalist production understood as a shift from a Fordist to a post-Fordist capitalist regime, and the emergence of new social movements and subjectivities—provided the political impetus for autonomist Marxists to reconsider both the changing political configuration of capitalism and its related means of domination, as well as the potential sites and subjectivities of resistance to that domination.[24]

Some of the main concepts that have been developed by Italian autonomist thinkers for apprehending the changing nature of capitalist production and control, and the political possibilities associated with the latter, are the concepts of *immaterial labour, general intellect*, and *the real subsumption of labour under capitalism*. Immaterial labour can be understood as

> the kind of labour that produces "the informational and cultural content of the commodity" ... [and] which, by its very nature, foregrounds the ability to activate and manage cooperation, in all of its affective, communicational and informational senses, for the sake of intensified productivity.[25]

21. Giuseppina Mecchia and Max Henninger, "Introduction: Italian Post-workerist Thought", *SubStance*, Vol. 36, No. 1 (2007), p. 3.

22. *Ibid.*

23. *Ibid.*, p. 4.

24. Alberto Toscano, "Factory, Territory, Metropolis, Empire", *Angelaki: Journal of the Theoretical Humanities*, Vol. 9, No. 2 (2004), pp. 200–201.

25. Alberto Toscano, "Vital Strategies: Maurizio Lazzarato and the Metaphysics of Contemporary Capitalism", *Theory, Culture & Society*, Vol. 24, No. 6 (2007), p. 73, citing Maurizio Lazzarato, "Immaterial Labor", in Virno and Hardt (eds.), *op. cit.*, p. 133.

Thus, under contemporary post-Fordist, global capitalism, the production of value increasingly depends on the properties of human being: language, communication, creativity, knowledge, affect, cooperation, etc. Note that while the concept was originally developed to describe both the technological changes—computerisation and informatisation—and the kind of labour associated with the informational or cultural worker,[26] Feminist thought has also called attention to the "other side" of immaterial labour—"affective labour"—that "covers a vast terrain of human interaction and trafficking, actual or virtual: health service, child care, domestic work, online dating, correspondence marriage, immigrant sex work, entertainment, tourism, digital information, and so on".[27]

Furthermore, and related to the previous point, under conditions of post-Fordist capitalism, the labour power that capital attempts to exploit is increasingly based on the human capacity for language, communication, and cooperation. For autonomist Marxists, this form of labour power is represented by the term *general intellect*.[28] The hypothesis is that the communicative and cooperative potential of human subjects is increasingly central to both capitalist production and reproduction.[29] Thus, the notion of general intellect highlights the fact that under contemporary capitalism "wealth is no longer the immediate work of the individual, but a general productivity of the social body—dispersed through technologies and human bodies, connected in new, shifting assemblages".[30] This, in turn, points towards the final key conceptual category, *the real subsumption of labour under capitalism*—again, a term introduced by Marx[31] and refashioned by autonomist Marxism. Initially theorised by Marx as a tendency within capitalism to subsume the social factors of production, autonomist Marxists have analysed contemporary capitalism as presenting conditions of an "actually existing real subsumption",[32] where increasingly

26. See Lazzarato, *op. cit.*

27. Sheldon H. Lu, *Chinese Modernity and Global Biopolitics: Studies in Literature and Visual Culture* (Honolulu: University of Hawai'i Press, 2007), p. 4. See also Michael Hardt, "Affective Labor", *Boundary 2*, Vol. 26, No. 2 (1999); Kathi Weeks, "Life within and against Work: Affective Labor, Feminist Critique, and Post-Fordist Politics", *Ephemera: Theory & Politics in Organization*, Vol. 7, No. 1 (2007).

28. The term is taken from the "Fragment on Machines" in Notebook VII of Marx's *Grundrisse*: "Nature builds no machines, no locomotives, railways, electric telegraphs, self-acting mules etc. These are products of human industry; natural material transformed into organs of the human will over nature, or of human participation in nature. They are organs of the human brain, created by the human hand; the power of knowledge, objectified. The development of fixed capital indicates to what degree general social knowledge has become a direct force of production, and to what degree, hence, the conditions of the process of social life itself have come under the control of the general intellect and been transformed in accordance with it". Karl Marx, *Grundrisse: Foundations of the Critique of Political Economy (Rough Draft)* (trans. Martin Nicolaus) (London: Penguin, 1973), p. 706.

29. Paolo Virno, "Notes on the General Intellect", in Saree Makdisi, Cesare Casarino and Rebecca E. Karl (eds.), *Marxism beyond Marxism* (New York: Routledge, 1996).

30. Tiziana Terranova, "Of Sense and Sensibility: Immaterial Labour in Open Systems", in Joasia Krysa (ed.), *Data Browser 03—Curating Immateriality: The Work of the Curator in the Age of Network Systems* (New York: Autonomedia, 2006), p. 29.

31. Marx contrasts the real subsumption of labour by capital with the merely formal subsumption in the first volume of *Capital*. He fleshed out in more detail the concepts of formal and real subsumption in the appendix to *Capital* titled "Results of the Immediate Process of Production". See Karl Marx, *Capital Volume 1* (trans. Ben Fowkes) (London: Penguin, 1990), pp. 1019–1038.

32. Jason Read, *The Micro-politics of Capital: Marx and the Prehistory of the Present* (Albany: State University of New York Press, 2003), p. 127.

> the creation of wealth no longer depends on working time narrowly defined, but coincides with the whole time of life ... it is the whole of social life—from child rearing to new forms of sexuality, from making music or videos on one's home computer to watching TV, from inventing new ways of dressing to making up a new way of speaking—that produces wealth.[33]

Finally, these concepts—immaterial labour, general intellect, and real subsumption—map out a new social and political terrain of command and control as well as contestation and resistance. It is in this context that the argument has been made that "*capitalism has become biopolitical*".[34] Of course, Hardt and Negri's argument regarding the biopolitical nature of production under global capitalism is by now well known. However, it is less commonly understood that this idea is not unique to Hardt and Negri (it is increasingly informing a number of critical studies into the changing nature of contemporary capitalism)[35]; nor is there any necessary connection between biopolitical capitalism and Hardt and Negri's claims regarding the new imperial nature of sovereignty and the immanence of the multitude to the new terrains of biopolitical production.[36]

It is also important to note that all of these concepts and their deployment within critical thought and political action are very much contested, not only by different forms of Marxian political thought but also within the Italian post-workerist tradition. Obviously this very brief discussion of some central elements of autonomist Marxist thought cannot do justice to the complexity of the theoretical and political issues involved or the debates they have generated within radical leftist thought. However, there are two implications regarding subjectivity and power than we can discern from this Italian "school" of Marxist though which might resonate with and help strengthen Foucauldian accounts of contemporary social and political life.

The first implication that can be drawn from autonomist theorisations of real subsumption under contemporary capitalism has to do with, on the one hand, the processes and mechanisms involved in subjectivation and, on the other, the nature of subjectivity itself. More specifically, one of the insights of autonomist thought is that under contemporary capitalist production and reproduction power acts on subjects in ways that work directly on pre- or sub-individual elements, such as the affective and libidinal structures of the subject.[37]

33. Terranova, *op. cit.*, 29.

34. Toscano, "Vital Strategies", p. 82; emphasis added.

35. Melinda Cooper, *Life as Surplus: Biotechnology and Capitalism in the Neoliberal Era* (Seattle: University of Washington Press, 2008); Lu, *op. cit.*; Majia Holmer Nadesan, *Governmentality, Biopower, and Everyday Life* (New York: Routledge, 2008); Read, *op. cit.*

36. Note that whilst for thinkers such as Negri and Lazzarato, conditions of real subsumption mark a biopolitical phase of capital, for Virno capitalism was biopolitical from the very beginning. See Paolo Virno, *A Grammar of the Multitude: For an Analysis of Contemporary Forms of Life* (Cambridge, MA and London: Semiotext (e), 2003), pp. 81–84. Note also that Negri distinguishes between the *biopower* of capital and the potential *biopolitics* of the multitude—a distinction that parallels the difference between *potenza* (potentiality) and *potere* (power). See Brett Neilson, "Potenza Nuda? Sovereignty, Biopolitics, Capitalism", *Contretemps*, Vol. 5 (2004).

37. Maurizio Lazzarato, "From Capital-labour to Capital-life", *Ephemera: Theory & Politics in Organization*, Vol. 4, No. 3 (2004); Virno, *A Grammar of the Multitude*, *op. cit.*, pp. 82–83. For a discussion of the "affective turn" in social theory in general, see the essays collected in Patricia Ticineto Clough and Jean

This can be seen both in the realms of work as well as consumption. In the former, Virno highlights the ways in which three dimensions of the contemporary worker—the self-employer, the professional, and the possessive individual—are produced through the mobilisation of affective dispositions and "linguistic-relational capacities".[38] Crucial for Virno is the fact that the nature of work under contemporary capitalism requires not so much a certain set of skills or knowledges but rather the cultivation of certain personal traits and dispositions, qualities—most importantly, the capacity for directing one's life as an enterprise—which are learned not in the spaces of work (or at least not primarily there) but in the extra-work spaces of social life. To put it differently, the place where the subject learns a relationship to itself that is homologous to the relationship between an entrepreneur and her business is not the workplace. Or it might be more accurate to say that the workplace is no longer confined to the factory or the office but is now diffused throughout social space.[39]

While Virno emphasises the fact that the contemporary working subject needs to be constructed, Lazzarato points out that the consumer is also not a naturally existing subject but also must be constructed. This also represents a shift in the dynamics of capitalism. Whereas under industrial capitalism the business created products and then tried to market and sell them, businesses nowadays are concerned less with the production and marketing of particular *products* than with the cultivation of a *brand*. For Lazzarato the company that best represents this shift is Benetton, whose publicity campaigns famously produce images not so much of commodities but of social subjects ("the AIDS patient, the newborn infant, the ship filled with Albanian refugees") whose relation to the company brand is left to the viewer's imagination.[40] What we are witnessing, Lazzarato argues, is that

> publicity does not serve merely to provide information about markets, but to constitute them. It enters into an "interactive" relationship with the consumer, addressing itself not only to her needs but above all to her desires. It addresses itself not only to her passions and emotions, but also directly interpellates "political" rationality. It produces not just the consumer but the "individual" of immaterial capitalism. It engages in dialogue with her convictions, values, and opinions, and has the courage to interpellate her where the political fears to go. Publicity is one of the most important forms of social and political communication at this turn of the century. Publicity as such increasingly occupies

O'Malley Halley (eds.), *The Affective Turn: Theorizing the Social* (Durham, NC: Duke University Press, 2007).

38. Paolo Virno, "Post-Fordist Semblance", *SubStance*, Vol. 36, No. 1 (2007), p. 45. See also Virno, *A Grammar of the Multitude*, *op. cit.*, pp. 84–93.

39. "Economy cannot be conceived within the time or place of the 'factory-office'. It has rather become spatially boundless and temporally endless: it is impossible to make the distinction between working time and free time, it is difficult to say where or when the actual act of production is being carried out, what is work and what is not, what creates value and what does not." Akseli Virtanen, "General Economy: The Entrance of Multitude into Production", *Ephemera: Theory & Politics in Organization*, Vol. 4, No. 3 (2004), p. 209.

40. Maurizio Lazzarato, "Strategies of the Political Entrepreneur", *SubStance*, Vol. 36, No. 1 (2007), p. 91.

> "public space," animates it, provokes it, arouses it. The enterprise produces "meaning" directly.[41]

Under these conditions of capitalist production of value, Lazzarato argues, the distinction between consumption and production become problematic since

> the actions of the consumer (her desires and values) are directly integrated, as a creative moment, into the social network of the enterprise ... [Marketing] constructs the product and solicits forms of subjectification. The consumer is no longer the passive mass-consumer of standardized commodities, but the active individual involved with the totality of her persona: to this end, it is necessary to "know" and solicit her ideology, lifestyle and conception of the world ... Capitalism is no longer the capitalism of production, but of the product. Marketing is no longer merely a technique for selling, but a mechanism that is constitutive of social relations, information and values for the market—one that integrates the techniques and "responsibility" of the political.[42]

It is not difficult to think of other examples where the "consumer" is put to work in the creation of value: from the participants in reality television shows to social networking websites such as MySpace and Facebook. In short, within this tendency towards immaterial labour, "our communication and our cultural practices are not only constitutive of social relations but are also a new form of labour increasingly integral to capital relations".[43] Furthermore, as the concept of enterprise becomes attached not only to businesses but also to the whole range of social entities, from the individual to the nation to the international organisation, so too is the concept of the brand and the concern for its management increasingly applied to all social actors.[44]

Additionally, it is important to keep in mind the crucial role of the different forms of communications media in the process of subjectivation described above. For Lazzarato, the subject who engages in the activity of consumption "is not reduced to the act of buying and carrying out a service or a product", but instead is incited to identify with and belong to a certain social imaginary. This imaginary, in turn, is constructed in part through advertising, which produces "a prompt to assume a form of living, i.e. a way of dressing, having a body, eating, communicating, residing, moving, having a gender, speaking, etc."[45]

41. *Ibid.*

42. *Ibid.*

43. In this regard, Deleuze's comment that "Marketing has become the center or 'soul' of the corporation" seems prescient. Gilles Deleuze, "Postscript on the Societies of Control", *October*, No. 59 (1992), p. 6; Mark Coté and Jennifer Pybus, "Learning to Immaterial Labour 2.0: Myspace and Social Networks", *Ephemera: Theory & Politics in Organization*, Vol. 7, No. 1 (2007), p. 92.

44. Paul du Gay, "Markets and Meanings: Re-imagining Organizational Life", in Majken Schultz, Mary Jo Hatch and Mogens Holten Larsen (eds), *The Expressive Organization Linking Identity, Reputation, and the Corporate Brand* (Oxford and New York: Oxford University Press, 2000). See also Miriam Salzer-Mörling and Lars Strannegård, "Silence of the Brands", *European Journal of Marketing*, Vol. 38, Nos. 1–2 (2004).

45. Maurizio Lazzarato, *Struggle, Event, Media* (2003), available: <http://www.republicart.net/disc/representations/lazzarato01_en.htm> (accessed 5 July 2009).

The other implication we can draw from autonomist theorisations of recent transformations in capitalism has to do with the functioning of power and its relation to subjectivity. With the notions of immaterial labour and real subsumption, we are already contemplating a shift from Fordist production where capitalist command is centred on the factory, to a post-Fordist form of capitalism where "labouring processes have moved outside of the factory to invest the entire society", producing, in effect, a "social factory".[46] Moreover, as capitalist production becomes more dispersed, likewise the accompanying mechanisms of control extend throughout the social body; thus regulation is increasingly "cultural" or, in Foucault's terminology, micropolitical. Here again, subjectivity is a crucial terrain for the operation of power. The *dispositifs* of power that have emerged with contemporary capitalism rely less on the techniques that Foucault described in terms of disciplinary power; rather, they operate increasingly through self-regulation, through "technologies of the self".[47]

However, it is important to keep in mind that the subjectivity that is regulated and made productive under contemporary capitalism is not the same form of subjectivity governed by the disciplinary powers of industrial capitalism. Or rather, to switch to a Deleuzian grammar, as capital becomes increasingly deterritorialised it works on subjects not so much through the external imposition of power but through a pressure given public expression in the imperative to be active, to communicate, to relate to others, which admonishes and materialises the subjective forms of immaterial labour. Thus, as Lazzarato argues, power is increasingly based on "the capacity to activate and manage productive cooperation"; we are exhorted to "become active subjects",[48] to communicate, to network, to constantly circulate and be visible within the "marketplace of ideas".

Furthermore, power seeks to manage the new productive subjectivities required by contemporary capitalism not only through the content of the various affective and cognitive elements of subjectivity but also through structuring the environment in which the subject operates—most importantly through the creation of a state of precarity for the working subject. In this context, "precarity" refers to "the growing insecurity brought on by the flexible management of the global work force within post-Fordist capitalism".[49] Thus precarity describes the general condition produced by the move towards the "flexibilisation" of the workforce, including part-time work, shift work, short-term contracts, self-employment, subcontracting, and the like.[50]

Actually, rather than distinguishing between, on the one hand, the mobilisation of bodily affects and the properties of subjectivity and, on the other, organisational and structural mechanisms of control, it might be more accurate to locate all

46. Michael Hardt and Antonio Negri, *Labor of Dionysus: A Critique of the State-form* (Minneapolis: University of Minnesota Press, 1994), p. 9.

47. Michel Foucault, "Technologies of the Self", in Luther H. Martin, Huck Gutman and Patrick H. Hutton (eds.), *Technologies of the Self: A Seminar with Michel Foucault*, (Amherst: University of Massachusetts Press, 1988).

48. Lazzarato, "Immaterial Labor", *op. cit.*, p. 135.

49. Enda Brophy, "System Error: Labour Precarity and Collective Organizing at Microsoft", *Canadian Journal of Communication*, Vol. 31, No. 3 (2006), p. 622.

50. Enda Brophy and Greig de Peuter, "Immaterial Labor, Precarity, and Recomposition", in Catherine McKercher and Vincent Mosco (eds.), *Knowledge Workers in the Information Society* (Lanham, MD: Lexington Books, 2007); Rosalind Gill and Andy Pratt, "In the Social Factory? Immaterial Labour, Precariousness and Cultural Work", *Theory, Culture & Society*, Vol. 25, Nos. 7–8 (2008).

these elements within new "diagrams of control" that have emerged within contemporary, post-Fordist capitalism.[51] This perspective would highlight how such diagrams work

> by reimposing centres and hierarchical distinctions against a much larger background of continuous variation (as the work on scale free networks demonstrates); by preemptively assigning objectives, outcomes and deadlines against the uneven temporality of processes of autonomous organisation which do not always follow their rhythm (as in the software industry); by channelling desire to prop up identities against the threat of dissipation (as in movements such as evangelical and nationalist blogs); by policing the rights of property against the indiscipline of non-linear circulation (as in the legal wars against peer-to-peer systems).[52]

As the above quote clearly indicates, this description of the mechanisms of control associated with contemporary capitalism is very much influenced by the thought of Gilles Deleuze, particularly his description of the transformation from the disciplinary societies analysed by Foucault to what Deleuze calls the new "societies of control".[53] Deleuze's theorisation of the new diagrams of power at work in the social field, as well as his critical analyses of the capitalist socius produced in collaboration with Félix Guattari,[54] have had an important influence on autonomist Marxist thought.[55] In some versions of autonomist thought, particularly Hardt and Negri's conceptualisation of the multitude, the deployment of Deleuze's thought (not to mention Foucault) can be problematic.[56] Nevertheless, what is striking is how, despite obvious differences, Deleuze's theorisation of the shift from disciplinary societies to societies of control and autonomist Marxist theorisations of the transformations associated with post-Fordist capitalism seem to converge in their descriptions of contemporary social and political reality.

What implications might this hold for Foucauldian analyses of world politics? Interestingly, in his 1979 lectures Foucault describes the subjectivities and political technologies of neoliberalism in a way that also resonates with many of the descriptions of contemporary capitalism discussed above. This is perhaps not surprising given the interconnection between the emerging neoliberalism that Foucault was concerned to understand in the late 1970s and the post-Fordist restructuring of capital that was the object of concern for autonomist Marxist thinkers. The following section, then, examines Foucault's analysis of neoliberalism, with an eye towards how this analysis can be productively employed to interrogate contemporary forms of governmentality—but also to see how it

51. Terranova, *op. cit.*, p. 33.

52. *Ibid.*

53. Deleuze, *op. cit.*

54. Gilles Deleuze and Félix Guattari, *Anti-Oedipus: Capitalism and Schizophrenia* (Minneapolis: University of Minnesota Press, 1983), Gilles Deleuze and Félix Guattari, *A Thousand Plateaus: Capitalism and Schizophrenia* (Minneapolis: University of Minnesota Press, 1987).

55. Hardt and Negri, *Empire, op. cit.*, pp. 27–32; Maurizio Lazzarato, "The Concepts of Life and the Living in the Societies of Control", in Martin Fuglsang and Bent Meier Sørensen (eds.), *Deleuze and the Social* (Edinburgh: Edinburgh University Press, 2006).

56. For a critical discussion, see Nicholas Thoburn, *Deleuze, Marx and Politics, Routledge Studies in Social and Political Thought*; 38 (London and New York: Routledge, 2003), pp. 69–102.

might open up a critical space for incorporating some of the insights into contemporary capitalism offered by autonomist Marxist thought.

Foucault on Neoliberal Subjectivity and Political Technologies

In the 1979 *Birth of Biopolitics*[57] Foucault's analysis took an interesting turn from the previous year's genealogy of liberal political rationality that he famously described in terms of "governmentality".[58] In what was for him a rare focus on contemporary political developments, Foucault provided an in-depth analysis of the rise of neoliberal political reason, first in post-Second World War Germany, then in the 1960s and 1970s in the United States. In his analysis of American neoliberalism, Foucault focuses on a key development in the idea of the economic subject of liberalism, or *Homo oeconomicus*. Under classical liberalism, Foucault argues, the liberal subject represented the limit to the sovereign's knowledge and power:

> [*Homo oeconomicus*] basically functions as what could be called an intangible element with regard to the exercise of power. Homo œconomicus is someone who pursues his own interest, and whose interest is such that it converges with the interest of others. From the point of view of a theory of government, homo œconomicus is the person who must be left alone. With regard to homo œconomicus, one must laisser-faire; he is the subject or object of laissez-faire.[59]

Foucault contrasts this with the neoliberal subject contemplated by contemporary American neoliberalism, as exemplified in the work of thinkers such as Gary S. Becker. The key development in a changing technology of government that Foucault identifies is the constitution of the neoliberal subject as a source of capital, as human capital.[60] With this invention, the subject exists no longer as one merely pursuing some interest, but rather as a potential source of capital that can and should work on itself in order to achieve the best return on its own investment; in American neoliberalism, "homo oeconomicus *is an entrepreneur, an entrepreneur of himself*".[61]

Furthermore, Foucault suggests, *Homo oeconomicus*

> is someone who accepts reality. Rational conduct is any conduct which is sensitive to modifications in the variables of the environment and which responds to this in a non-random way, in a systematic way, and economics can therefore be defined as the science of the systematic nature of responses to environmental variables.[62]

57. Michel Foucault, *The Birth of Biopolitics: Lectures at the Collège de France, 1978–79* (ed. Arnold I. Davidson; trans. Graham Burchell) (New York: Palgrave Macmillan, 2008).

58. Michel Foucault, *Security, Territory, Population: Lectures at the Collège de France, 1977–1978* (ed. Arnold I. Davidson; trans. Graham Burchell) (New York: Palgrave Macmillan, 2007).

59. Foucault, *Birth of Biopolitics, op. cit.*, p. 270.

60. See, for example, Gary S. Becker, *The Economic Approach to Human Behavior* (Chicago: University of Chicago Press, 1976); *idem*, *Human Capital: A Theoretical and Empirical Analysis, with Special Reference to Education*, 2nd edn (New York: Columbia University Press, 1975).

61. Foucault, *Birth of Biopolitics, op. cit.*, p. 226; emphasis added.

62. *Ibid.*, p. 269.

This aspect of the neoliberal subject—the neoliberal subject as "the person who accepts reality or who responds systematically to modifications in the variables in the environment"—makes possible a profound transformation in the art of government, for this new neoliberal subject "appears precisely as someone manageable ... [as] someone who is eminently governable".[63] The consequence, Foucault argues, is that "[f]rom being the intangible partner of *laissez-faire, homo oeconomicus* now becomes the correlate of a governmentality which will act on the environment and systematically modify its variables".[64] In other words, this new neoliberal subject makes possible a transformation in the art of government; government becomes both more indirect and also more extensive to the social field. Foucault's conclusion regarding this new art of government is profoundly suggestive:

> [Y]ou can see that what appears on the horizon of this kind of analysis is not at all the ideal or project of an exhaustively disciplinary society in which the legal network hemming in individuals is taken over and extended internally by, let's say, normative mechanisms. Nor is it a society in which a mechanism of general normalization and the exclusion of those who cannot be normalized is needed. On the horizon of this analysis we see instead the image, idea, or theme-program of a society in which there is an optimization of systems of difference, in which the field is left open to fluctuating processes, in which minority individuals and practices are tolerated, in which action is brought to bear on the rules of the game rather than the players, and finally in which there is an environmental type of intervention instead of the subjugation of individuals.[65]

In the written manuscript for the lecture quoted above, Foucault elaborates further the new configuration of power as it relates to what he describes in terms of a new "human technology". Here, Foucault identifies

> a massive withdrawal with regard to the normative-disciplinary system. The correlate of the system formed by a capitalist type of economy and political institutions indexed to the law was a technology of human behavior, an "individualizing" governmentality comprising: disciplinary control (*quadrillage*), unlimited regulation, subordination/classification, the norm. Considered overall, liberal governmentality was both legalistic and normalizing, disciplinary regulation being the switch-point between the two aspects ... [complicated by] the ultimate incompatibility between legal forms and normalization. This system no longer seems to be indispensable. Why? Because the great idea that the law was the principle of governmental frugality turns out to be inadequate ...[66]

Foucault continues by observing that within neoliberal governmentality the law cannot function as the ultimate principle of governmental rationality. The law,

63. *Ibid.*, p. 270.
64. *Ibid.*, pp. 270–271.
65. *Ibid.*, pp. 259–260.
66. *Ibid.*, p. 260.

therefore, needs to be reconceptualised; it needs to be understood as having a distinct form and function. The form is prohibition and constraint. The function is to establish the rules of the game in order to "*[enable] everybody to be a rational subject, i.e., to maximize the functions of utility*".[67] This then raises the problem, Foucault suggests, of "[h]ow to remain within [a system of the] Rule of law? How to rationalize this enforcement, it being understood that the law itself cannot be a principle of rationalization?" The answer that neoliberal thought comes up with, Foucault suggests, is

> through the calculation of costs, the utility of the law and the cost of its enforcement, and by the fact that if you do not want to get out of the law and you do not want to divert its true function as rule of the game, the technology to be employed is not discipline-normalization but action on the environment. Modifying the terms of the game, not the players' mentality.[68]

Remarkably, Foucault here seems to anticipate what Deleuze a decade later will describe as the shift from disciplinary society to a society of control, an account of transformations in the operations of power that, as was mentioned above, has had a major impact on many autonomist Marxist thinkers. In fact, the new functioning of the apparatuses of power associated with neoliberalism has been studied by some of those who collaborated with Foucault. Thus, for example, Robert Castel describes the replacement of a concrete *subject* that is the object of government with "a combinatory of *factors*, factors of risk".[69]

> New forms of control are appearing . . . which work neither through repression nor through . . . welfare interventionism . . . In place of these older practices, or rather alongside them, we are witnessing the development of differential modes of treatment of populations, which aim to maximize the returns on doing what is profitable and to marginalize the unprofitable. Instead of segregating and eliminating undesirable elements from the social body, or reintegrating them more or less forcibly through corrective or therapeutic interventions, the emerging tendency is to assign different social destinies to individuals in line with their varying capacity to live up to the requirements of competitiveness and profitability.[70]

Clearly, the picture that emerges from this theorisation of the shift from disciplinary society to a society of control involves a complex relationship between mechanisms of control and subjectivity. At the end of the section on human technology, we can see Foucault grapple with this problem. At the very end of the text, after he discusses the argument made by American neoliberalism in favour of regulation through environmental techniques—especially the manipulation of supplies and demands—Foucault asks: "But does this mean that we are dealing with natural subjects?"[71] In other words, what are the qualities specific to the subject that

67. *Ibid.*, emphasis added.
68. *Ibid.*
69. Robert Castel, "From Dangerousness to Risk", in Burchell *et al.* (eds.), *op. cit.*, p. 281.
70. *Ibid.*, p. 294.
71. Foucault, *Birth of Biopolitics*, *op. cit.*, p. 261.

neoliberalism must assume to exist in order to formulate technologies which aim to govern the subject? Or, alternatively, is it that rather than starting from a given conceptualisation of the subject, neoliberalism institutes a "fictional" subject—fictional in the sense that there does not exist a natural subject to be worked upon, but rather the subject is the result of a combination of factors, including the governmental technologies proposed by neoliberal thought.

Whether Foucault himself reached the latter conclusion, this would seem the correct one; the neoliberal subject is not a natural subject, at least not the natural subject of classical liberalism. The constitution of calculation as a comprehensive grid of intelligibility for all human action and interaction does away with the need for the kind of "anthropological categories and frameworks developed by the human and social sciences".[72] The only characteristic that is ascribed to human subjects is the ability to make choices that take into account the reality in which the subject finds herself. This is the subject of rational choice. However, in the writings of Becker and other neoliberal thinkers, reason is understood not as a universal category that could under ideal circumstances be determined by, for example, determining whether the optimal choice was made with regard to the desired outcome and the given conditions. The quality of the decision will vary according to the value of the human capital available; behaving rationally is defined in a minimal sense as acting in a non-random way in light of a given reality. "*Homo oeconomicus* is someone who accepts reality."[73] Or, as Becker writes, "Even irrational decision units must accept reality and could not, for example, maintain a choice that was no longer within their opportunity set. And these sets are not fixed or dominated by erratic variation, but are systematically changed by different economic variables [. . .]."[74]

What is crucial, however, for neoliberal governmentality is that the subject not only exercises her capacity for rational choice but that this choice is made within a social setting where she alone is responsible for, and bears the consequences of, the outcomes of that decision. This is made clear in Becker's recent discussion of the attempt by some to shift blame for the so-called "sub-prime" loans from the borrowers to the lenders. Becker is quite explicit that individual responsibility is a crucial foundation for a liberal, market society:

> Successful attempts to shift the responsibility for bad decisions toward others and to society more generally create a "moral hazard" in behavior. If individuals are not held accountable for decisions and actions that harm themselves or others, they have less incentive to act responsibly in the first place since they will escape some or all of the bad consequences of their actions. It does not matter greatly whether this moral hazard resulted from the shifting of blame for unsuccessful actions to the "small print"

72. Colin Gordon, "Governmental Rationality: An Introduction", in Burchell *et al.* (eds.), *op. cit.*, p. 43. This argument is made most forcefully in Becker, *op. cit.*, p. 8: "[t]he economic approach is a comprehensive one that is applicable to *all human behavior*, be it behavior involving money prices, or imputed shadow prices, repeated or infrequent decisions, large or minor decisions, emotional or mechanical ends, rich or poor persons, men or women, adults or children, brilliant or stupid persons, patients or therapists, businessmen or politicians, teachers or students".

73. Foucault, *Birth of Biopolitics*, *op. cit.*, p. 269.

74. Gary S. Becker, "Irrational Action and Economic Theory", *Journal of Political Economy*, Vol. 70, No. 4 (1962), p. 167, quoted in Foucault, *Birth of Biopolitics*, *op. cit.*, p. 287.

> in a contract, to an abused childhood, to a mental state, or to many other efforts to shift responsibility away from oneself.
>
> An important foundation of the philosophy behind the arguments for private enterprise, free economies, and free societies more generally, is that these societies rely on and require individual decision-making and responsibility. This philosophy not only emphasizes the moral hazard reasons to require individual responsibility, but also "the use it or lose it principle", a colloquial expression indicating that various mental and physical capacities wear down and erode if they are not used on a regular basis. This principle implies that people who are accustomed to having other persons or governments make their decisions for them lose the ability to make good decisions for themselves. Free societies lead to better decision-making partly because men and women accumulate more experience at making decisions that affect their well-being and that of others.[75]

We can see here a profound ambivalence within neoliberalism. On the one hand, human nature is assumed to be that of the rational, calculating subject. On the other, there is also the fear that this natural capacity needs to be instilled, fostered, and protected against outside forces that could erode it, such as paternal protection from the state: with regard to human capital, one must "use it or lose it".[76] So much of neoliberal governance is about creating the conditions that encourage, and indeed make possible, the rational, calculating subject. Here we can find an imperative for government—to make sure that individuals are responsible for their actions, that they are rewarded for good decisions and punished for bad ones. Indeed, as Lazzarato argues,

> Liberalism is first and foremost neither an economic theory nor a political theory; it is rather an art of government that assumes the market as the test and means of intelligibility, as the truth and the measure of society ... By market we do not mean "commodification". According to Foucault ... the market is not defined by the human instinct to exchange ... by market we must always understand competition and inequality, rather than equality of exchange. Here, the subjects are not merchants but entrepreneurs. The market is therefore the market of enterprises and of their differential and non-egalitarian logic.[77]

Finally, with regard to neoliberal subjectivity, the idea that the neoliberal subject is defined as one who "accepts reality" requires further elaboration. Foucault points out the fundamental importance of this understanding for the development of new technologies of government, ones that Deleuze described as mechanisms of control. However, more needs to be said about this "reality" that must be accepted by the neoliberal subject. As the analysis presented here has already

75. Gary S. Becker, *The Erosion of Individual Responsibility* (2008), available: <http://www.becker-posner-blog.com/archives/2008/03/the_erosion_of.html> (accessed 16 June 2009). I would like to thank Nicholas Kiersey for calling this article to my attention.

76. *Ibid.*

77. Maurizio Lazzarato, *Biopolitics/Bioeconomics: A Politics of Multiplicity* (n.d.), available: <http://www.generation-online.org/p/fplazzarato2.htm> (accessed 10 June 2008).

suggested, perhaps the most important way in which neoliberalism shapes subjectivity is in suggesting that each individual is the bearer of a human capital (however this concept is ultimately worded), who must seek to maximise her own self-value, and as a consequence must take into account the "rules of the game".

Leaving aside, for the moment, the question of how the production of this subject is achieved, we can already see that these three elements of neoliberal subjectivity make possible a number of ways for governmental techniques to direct this subject's conduct. In short, this foundation for subjectivity means that government does not need to affect individual subjects in a deep sense,[78] at least in the sense that Foucault speaks of disciplinary power reaching into the subject's "soul", but rather through motivating the subject to perform or act in certain ways. In other words, the subject does not need to internalise specific neoliberal values about the importance of a society constituted by competitive, rationally choosing individuals, provided she accepts that this is the reality in which she finds herself and acts accordingly. Thus, far from being naturally occurring phenomena, markets and rationally calculating subjects are "what liberal government must make possible and real ... [T]he mechanisms of the market (prices, laws of demand and supply) are fragile. Favourable conditions must be continuously created for these fragile mechanisms to function."[79]

As already suggested, one of the ways in which government makes possible the functioning of the market, of the action of supply and demand as a social regulator, is through the production of the competitive subject. Thus one potentially powerful form of governmental intervention is that of directing subjects by suggesting the general contours of how one can/should work towards the ultimate goal of the maximisation of one's human capital. Thus, techniques such as critical evaluations made in reference to a standard—i.e. benchmarking, but also forms of self-benchmarking—might work not necessarily through a process whereby the subject internalises the idea that benchmarking or self-evaluation is good per se, but rather through convincing the subject that this is the way the game is currently played. In this way, subjects might reproduce governmental techniques such as benchmarking, not through any internalisation of a norm regarding the value of benchmarking but as a strategic action based on accepting the idea of benchmarking as a norm—i.e. as one of the ways in which "the game" is "played". If this is the case, it would then be important to take into account that subjects often do act strategically, and that certain technologies and rationalities get reproduced not because of any underlying ethos that has been internalised but through subjects acting strategically. This, however, would not mean that we have to understand the subject as being "naturally" calculative; quite the contrary, it merely would make the relative success of the constitution of the rationally calculative subject an important, indeed crucial, part of the analysis.

This is where the notion of performativity and embodied subjectivity might be helpful for apprehending the points of contact between subjectivity and neoliberal technologies and rationalities of government. The production of neoliberal subjectivity requires not only that subjects internalise the notion that they are engaged in an enterprise of maximising their self-value, in the context of an always already

78. Although formulating the subject as a self-entrepreneur who must maximise her own value by taking into account the way the game in which she operates is structured is already quite deep.

79. Lazzarato, *Biopolitics/Bioeconomics*, *op. cit.* See also U. Kalpagam, "Colonial Governmentality and the 'Economy'", *Economy and Society*, Vol. 29, No. 3 (2000).

structured game with its own rules; it also requires that the rules of the game are embodied and performed. The embodied affects, dispositions, and performances of neoliberal subjects, in turn, are shaped by a wide array of media that seek to provide the subject with information—advice and guidelines (i.e. the rules of the game)—about how to successfully *be* an entrepreneur of oneself. In other words, crucial for the production of neoliberal subjects are what Foucault termed "technologies of the self".[80] And, as Lazzarato points out, for Foucault, these neoliberal subjects

> operate and constitute the real by the enactment and the addition of bits, pieces, parts each time singular. The "truth" of these parts and these bits and pieces cannot be found in the political or the economic "whole". Through the market and society the art of government is deployed with an increasing capacity of intervention, intelligibility and organisation of the whole of juridical, economic and social relations from the standpoint of the entrepreneurial logic.[81]

This description of governmental technologies that both help constitute the rational choosing subject and regulate that subject through modifications in her environment—by "systematically" changing the subject's "opportunity set" through "different economic variables"—clearly resonates with both autonomist Marxist descriptions of the functioning of power mechanisms under contemporary capitalism, as well as Deleuze's description of societies of control. There is clearly an opportunity, then, for governmentality studies to incorporate some insights gleaned from autonomist Marxist thought into their analyses of contemporary forms of neoliberal governmentality. This, at least, has been the aim of this paper: to introduce autonomist Marxist thought to a broader IR audience and hopefully to open up a new critical space for thought. The goal here is not to provide a detailed blueprint for moving governmentality studies in a new direction, but rather to suggest possible future lines of research.

Conclusion: Bringing Capitalism Back in

The governmentality approach to studying social and political life is perhaps best understood not as a full-blown theory but rather as a certain methodology for approaching a specific problem-space, namely: how, on the basis of what rationalities and through what kinds of techniques and practices are subjects governed?[82] For the most part, governmentality scholars have focused on neoliberal forms of governmentality. In their analyses of neoliberalism, governmentality studies have avoided what are seen as a problematic reduction of neoliberalism to the interests of an easily identifiable class or group of institutions, as might be seen in some forms of Marxian analysis. The understanding of neoliberalism within governmentality scholarship is based on the assumption that whatever connection may exist between, on the one hand, particular rationalities and techniques of government and, on the other hand, capitalist modes and relations of production, we

80. Foucault, "Technologies of the Self", *op. cit.*

81. Lazzarato, *Biopolitics/Bioeconomics*, *op. cit.*

82. However, see Merlingen, "Foucault and World Politics", *op. cit.*, for an argument in favour of describing governmentality in terms of a theory.

cannot deduce the ways in which the former are formed and operate from the latter. In this regard, governmentality scholars have followed, in a certain sense, Foucault who detailed the emergence of disciplinary forms of power that, while clearly interconnected with capitalist production and social relations, could not be explained solely in those terms:

> [Capitalism], as it was established in the nineteenth century, was obliged to elaborate a set of political techniques, techniques of power, by which man was tied to something like labor—a set of techniques by which people's bodies and their time would become labor power and labor time so as to be effectively used and thereby transformed into hyperprofit. But in order for there to be hyperprofit, there had to be an infrapower. A web of microscopic, capillary political power had to be established at the level of man's very existence, attaching men to the production apparatus, while making them into agents of production, into workers.[83]

Thus, Foucault's work has often been viewed as "a radical departure from or even as a critique of many of the methodological and political assumptions" "of both Marxist analysis and socialist strategy".[84]

However, it is important to keep in mind the intellectual and political context in which Foucault was working—one where leftist thought was dominated by a certain form of Marxism which aspired to provide a general theory of social and political reality in which capitalism functioned as an overall explanatory grid for all historical developments. At least this is the way Foucault has described the context for his complex and ambiguous relationship with Marxist thought,[85] leaving aside the accuracy of this portrayal of Marxism in the 1960s and 1970s. Coincidentally, some of the concerns and critiques Foucault had of Marxist thought were shared by many of the post-workerist and autonomist thinkers discussed here. Furthermore, these thinkers have found Foucault's concept of biopolitics to offer a powerful analytical lens for apprehending recent transformations in global capitalism. What is more, a number of elements found within Foucault's analysis of contemporary neoliberalism seem to parallel many of the descriptions of the forms of power and subjectivity immanent to global capitalism as described by autonomist Marxist thought. Finally, governmentality as a general approach to political analysis has proved flexible enough to develop in tandem with other compatible theoretical approaches such as actor network theory.[86]

83. Michel Foucault, "Truth and Juridical Forms", in James D. Faubion (ed.), *Essential Works of Foucault 1954–1984*, Vol. 3: *Power* (New York: New Press, 2000), p. 86. See also Michel Foucault, *The History of Sexuality: An Introduction* (New York: Vintage Books, 1990), p. 141.

84. Barry Smart, "The Politics of Truth and the Problem of Hegemony", in Barry Smart (ed.), *Michel Foucault: Critical Assessments* (London and New York: Routledge, 1994), p. 208.

85. See, for example, Michel Foucault, "Critical Theory/Intellectual History", in Lawrence D. Kritzman (ed.), *Politics, Philosophy, Culture: Interviews and Other Writings, 1977–1984* (New York: Routledge, 1988); *idem*, *Remarks on Marx: Conversations with Duccio Trombadori* (trans. R.J. Goldstein and J. Cascaito) (New York: Semiotext(e), 1991). See also Foucault's discussions with his Marxist interlocutors in Michel Foucault, *Power/Knowledge: Selected Interviews and Other Writings, 1972–1977* (Brighton: Harvester Press, 1980).

86. See, for example, Gavin Kendall, "Global Networks, International Networks, Actor Networks", in Wendy Larner and William Walters (eds.), *Global Governmentality: Governing International Spaces* (London and New York: Routledge, 2004).

As I see it, there are two possible ways in which governmentality approaches could take on board the treatments of contemporary capitalism presented by autonomist Marxist thought. One way would be to integrate a Foucauldian approach within a more general Marxist political problematic of a critical investigation of capitalism, as scholars such as Jan Selby have called for.[87] I find this move problematic, however. To begin with, contrary to what some have argued, the idea of incorporating Foucauldian analyses of the "how" of power into broader Marxist questions of the "why" of power ("the ceaseless accumulation of capital, and attendant conflicts amongst capitalists, classes and states")[88] is premised on a problematic assumption: namely that there necessarily exists an objective "why" of power that can be discerned a priori and independent of the analysis we carry out. For Foucauldian scholars who find value in the French philosopher's historical nominalism, the idea that capitalism predetermines the "why" of power is unlikely to be very convincing.

Just as problematic is the idea that Foucauldian analyses can offer fruitful investigations of the "ascending" forms of micropower, while Marxist theory can usefully detail "how power also 'descends' from the state in the form of interests, strategies and decisions".[89] This ignores Foucault's argument that rather than "taking as a primary, original, and already given object notions such as the sovereign, sovereignty, the people, subjects, the state, and civil society" we should start with the different discursive and material practices that both make it possible to speak of "the state" and also for the state to achieve a degree of discursive and material "reality".[90] In fact, it is precisely in part a response to this methodological challenge, Foucault argues, that he felt the need to invent the concept of governmentality—a notion, he stresses, that is meant to serve as an analytical grid for examining a diverse array of power relations regardless of the particular site or scale. Thus, he argues,

> the analysis of micro-powers, or of procedures of governmentality, is not confined by definition to a precise domain determined by a sector or the scale, but should be considered simply as a point of view, a method of decipherment which may be valid for the whole scale, whatever its size.[91]

Or, as Lazzarato argues, "Liberal macro-governmentality is only possible because it exerts its micropowers upon a multiplicity. These two levels are inseparable."[92]

In fact, herein lies one of the strengths of a governmentality approach—the way it "brackets" "the world of underlying forces and causes, and instead [examines]

87. Selby, *op. cit.*

88. *Ibid.*, p. 339.

89. *Ibid.*, pp. 339–340.

90. Foucault, *Birth of Biopolitics*, *op. cit.*, pp. 2–3. In a later lecture Foucault claims: "the state does not have an essence. The state is not a universal nor in itself an autonomous source of power. The state is nothing else but the effect, the profile, the mobile shape of a perpetual statification (*étatisation*)". *Ibid.*, p. 77.

91. *Ibid.*, p. 186. While Foucault here speaks of scale, his methodological argument resonates with the proposal by some human geographers to abandon the notion of scale in favour of a "flat ontology". See Sallie A. Marston, John Paul Jones III and Keith Woodward, "Human Geography without Scale", *Transactions of the Institute of British Geographers*, Vol. 30, No. 4 (2005); *idem*, "Situating Flatness", *Transactions of the Institute of British Geographers*, Vol. 32, No. 2 (2007).

92. Lazzarato, *Biopolitics/Bioeconomics*, *op. cit.*

the different ways in which the real has been inscribed in thought". Thus, for example, rather than taking for granted the existence of something called "the global" that forms the basis for ideas such as globalisation, a governmentality approach instead views globalisation as "one particular political imagination amongst many, rather than the underlying logic of an epoch or the outcome of global pressures such as international competitiveness".[93] This means that discourses of the global—globalisation, global governance, global civil society—are thus seen as a particular form of power/knowledge that makes possible and also forecloses different kinds of political practices and arrangements.[94]

For a number of reasons, these incompatibilities would unlikely be resolved if the type of Marxist framework into which a governmentality perspective were to be integrated is autonomist rather than, say, neo-Gramscian. Trying to work within such overly broad conceptual categories as Empire and multitude would likely strip a governmentality of much of what it has to offer to social and political analysis, particularly its nominalist methodology just discussed. For this reason, a second possible way for governmentality studies to take seriously autonomist Marxism's analysis of contemporary capitalism seems preferable. This would consist of a critical engagement with many of the theoretical claims made by autonomist Marxist thinkers regarding the changing nature of capitalism and related changes in the operations of power and its relation to subjectivity. This need not take the form of a wholesale importation of the conceptual apparatus of autonomist Marxism. Indeed, as Nigel Thrift argues, terms such as "immaterial labour" are dispensable;[95] the phenomena and trends they point towards—for example the way in which capital increasingly seeks to draw on and make productive the various components of human subjectivity—are areas that easily lend themselves to investigation from a governmentality framework. Indeed, there are already a number of scholars in diverse disciplines who draw inspiration from Foucault and autonomist Marxist thought—as well as from other theoretical traditions and schools of thought.[96]

In conclusion, what is being proposed here is to widen the space of inquiry to which governmentality studies are directed. So far this research programme has produced a number of richly empirical studies, in a range of disciplines and fields of study, of the different governmental projects that seek to shape and direct social reality. However, if we consider the basic notion of governmentality—the conduct of conduct and the structuring of the field of action for possible conduct—then by focusing exclusively on neoliberal forms of governmentality we leave out a wide range of social and political phenomena related to the global expansion of capital that are directly concerned with governing subjects and

93. Larner and Walters, "Globalization as Governmentality", *op. cit.*, p. 16.

94. *Ibid.* The idea that space is not a natural, but rather a socially constructed, phenomenon is not unique to a Foucauldian perspective. See, for example, Henri Lefebvre, *The Production of Space* (Oxford, UK and Cambridge, MA: Blackwell, 1991); Edward W. Soja, *Postmodern Geographies: The Reassertion of Space in Critical Social Theory* (London and New York: Verso, 1989). For a critical genealogy of the global, see Denis E. Cosgrove, *Apollo's Eye: A Cartographic Genealogy of the Earth in the Western Imagination* (Baltimore: Johns Hopkins University Press, 2001).

95. Nigel Thrift, *Non-representational Theory: Space, Politics, Affect* (New York: Routledge, 2007), p. 48.

96. Cooper, *op. cit.*; Nadesan, *op. cit.*; Jeffrey T. Nealon, *Foucault beyond Foucault: Power and its Intensifications since 1984* (Stanford: Stanford University Press, 2008); Jason Read, "A Genealogy of Homo-Economicus: Neoliberalism and the Production of Subjectivity", *Foucault Studies*, No. 6 (2009); *idem, Micro-politics of Capital.*

their relations today. There are obviously a wide range of theoretical resources available for expanding the domain of analysis of governmentality to the contemporary forms of capitalist power and subjectivity, a discussion of which is beyond the scope of this paper.[97] The aim of this paper has been to open up a critical engagement with one such theoretical body—autonomist Marxist thought—and hopefully provoke governmentality scholars to push their analysis in new directions.

97. Ash Amin and Nigel J. Thrift, *The Blackwell Cultural Economy Reader* (Malden, MA: Blackwell, 2003); Jean-François Bayart, *Global Subjects: A Political Critique of Globalization* (trans. Andrew Brown) (Cambridge: Polity, 2008); Richard Harvey Brown (ed.), *The Politics of Selfhood: Bodies and Identities in Global Capitalism* (Minneapolis: University of Minnesota Press, 2003); Jean Comaroff and John L. Comaroff, *Millennial Capitalism and the Culture of Neoliberalism* (Durham, NC: Duke University Press, 2001); J.K. Gibson-Graham, *The End of Capitalism (as We Knew It): A Feminist Critique of Political Economy*, 2nd edn (Minneapolis: University of Minnesota Press, 2006); Nigel Thrift, *Knowing Capitalism* (London: Sage, 2005).

Governmentality of What? Populations, States and International Organisations

JONATHAN JOSEPH*

As more work on governmentality appears in International Relations (IR), it is time to take stock and deal with a few questions. In social theory, the governmentality approach has mainly addressed "advanced liberal" societies and can be defined as having the health, wealth and well-being of populations as its target and governance from a distance through the "conduct of conduct" as its means of operating. There are two major problems in transferring governmentality to IR. First, not all societies can be described as "advanced liberal" ones. Second is the problem of whether there is such a thing as "global governmentality". This article argues that the lack of the necessary social conditions does indeed make it difficult to apply the technologies of governmentality to various parts of the world. However, the aim of international organisations might be less the regulation of populations as the application of governmentality to states.

Introduction

Some years after everyone else in the social sciences, International Relations (IR) theorists have been discovering the importance of governmentality. This means a focus on the different institutions, procedures, analyses and reflections that target the population by trying to shape people's conduct.[1] The governmentality approach has worked well in a number of areas of social theory such as insurance,[2] accounting,[3] crime[4] and health,[5] bringing a degree of subtlety and sophistication to the study of practices of governance in advanced liberal societies. A recent flurry of new work in the field of IR invites assessment as to whether this approach is able to work as effectively in assessing international politics as it does in looking at interventions in domestic politics. Indeed, more than this,

---

*The author would like to thank Ruth Blakeley, Nicholas Kiersey and Jan Selby for their comments.

1. Michel Foucault, *Security, Territory, Population* (Basingstoke: Palgrave, 2007), p. 108.

2. See the chapters by Jacques Donzelot and François Ewald in Graham Burchell, Colin Gordon and Peter Miller (eds.), *The Foucault Effect: Studies in Governmentality* (Chicago, IL: University of Chicago Press, 1991).

3. Michael Power, *The Audit Society: Rituals of Verification* (Oxford: Oxford University Press, 1997).

4. Pat O'Malley, "Risk, Power and Crime Prevention", *Economy and Society*, Vol. 21, No. 3 (1992), pp. 252–275.

5. Peter Miller and Nikolas Rose, *The Power of Psychiatry* (Cambridge: Polity, 1986); Thomas Osborne, "Liberalism, Neo-liberalism and the Liberal Profession of Medicine", *Economy and Society*, Vol. 22, No. 3, pp. 345–356.

we have to assess the theoretical question of how adequately the *theory* of governmentality explains developments in IR, but also, depending on the answer to this question, how adequately governmentality itself works in *practice*. These areas of practice are fairly diverse. They range from the operations of private security companies to the activities of non-governmental organisations (NGOs), to the interventions of international organisations and the issue of global governance (whether the relations between states, intergovernmental organisations, international law, NGOs, private-sector actors and those in civil society can be considered to provide a basis for a global system of regulation). This raises important questions about governmentality both as a concept and a practice. An important theoretical issue concerns the difference between using the governmentality approach to understand what goes on at the local level in different places around the world (a comparative approach to different forms of governmentality across the globe) and developing the idea that there is something substantially different operating at a higher level that deserves the name global governmentality. Ontological questions should address the differences in forms of governmentality operating both in different places and at various levels. For example, why are the techniques of governmentality effective in some places but not in others? What does this say about its conditions of possibility? And crucially for this piece, if we distinguish between the workings of governmentality in different parts of the world and the idea of global governmentality, then if there is such a thing as the latter, what is its object (populations, states, or institutions?) and who are its agents? This article will open the discussion by considering what governmentality is and examining how it relates to sovereignty and disciplinary power. It argues that governmentality is distinguished by its concern for the population and its liberal way of doing this. When looking at contemporary politics, the dominant form of governmentality in advanced liberal societies is a neoliberal governmentality that promotes the idea of freedom through the encouragement of competition. While IR theory has recently shown itself willing to take up the idea of governmentality and apply it to global developments, very few, if any, of the theorists discussed below have been willing to address the question of whether this sort of neoliberalism can be forced on non-liberal societies. The second section questions whether governmentality really can work outside of those societies that can be characterised as having advanced liberal rule.[6] The third section asks whether there can be such a thing as global governmentality operating according to neoliberal principles. The conclusion will try to square some of the contradictions.

What is Governmentality?

Michel Foucault's work on governmentality does not emerge in any systematic form; rather, it is developed in a series of lectures given at the Collège de

6. The term "advanced liberalism" is used by Miller and Rose to characterise those societies with such things as multiple social technologies, new specification of the subject of governance and a new relation between expertise and politics. Peter Miller and Nikolas Rose, *Governing the Present* (Cambridge: Polity, 2008), pp. 18, 212–213. They prefer to talk of "advanced liberalism" as a form of governmentality rather than as a type of society. Here the term will be made more ontological to apply to types of societies where this type of governmentality works owing to the particular nature of their capitalist social development and the institutional changes of the last part of the 20th century.

France which have subsequently been transcribed and published. In these lectures, Foucault is still clearly thinking through the concept and applying it in a number of different ways in order to examine a range of different questions. As a result we can safely say that there are a number of very general definitions of what governmentality is; but otherwise, the concept is applied in different ways in different contexts. This means that there is no exact understanding of what governmentality is; we have to see it as belonging to a field of analysis where it is deployed alongside other concepts and ideas with which it intersects and engages. Even the most general definition of governmentality as the "conduct of conduct" (*conduire des conduites*)[7] requires considerable investigation. This definition suggests that governance takes place from a distance as the power to influence the actions of others. To understand this further, we have to see how government forms part of a triangle of sovereignty–discipline–government and where it stands out as having the population as its target.[8]

Foucault argues that the problem of government starts to emerge in the 16th century and gradually begins to break free of the constraints of sovereign power (as exercised over a territory and its inhabitants[9]) in the 18th century. The problem of government is concerned with populations and their conduct and starts to be addressed by means of the introduction of political economy into their practices.[10] This works by introducing laissez-faire notions of freedom of conduct derived from the "natural processes" of the economic sphere.[11] Consequently, government comes to be understood as respecting the freedom of such processes through the deliberate self-limiting of government—something that is considered an intrinsic part of governmental rationality.[12] This can be contrasted with sovereign power with its concern for territory and disciplinary power which functions in a more coercive and preventive way.[13] Foucault examines the ways in which government works from a distance, employing new techniques of observation, calculation and administration, going beyond any definite limits of state power to express itself though an ensemble of "institutions, procedures, analyses, and reflections, the calculations and tactics that allow the exercise of this very specific, albeit very complex, power".[14] This does not mean a rejection of sovereignty or state power but quite the opposite. However, it requires a new way of thinking about state and sovereign power alongside a range of other institutions and practices. It also means a shift to the micro level so that "rather than asking ourselves what the sovereign looks like from on high, we should be trying to discover how multiple bodies, forces, energies, matters, desires, thoughts and so on are gradually, progressively, actually and materially constituted as subjects".[15] Consequently, the state is "superstructural" in the sense that it colonises, uses and transforms already existing micropowers which can then be redeployed by

7. Michel Foucault, *Dits et écrits IV* (Paris: Gallimard, 1994), p. 237. This term is not translated in this way in English versions of Foucault.

8. Foucault, *Security, Territory, Population*, *op. cit.*, p. 108.

9. *Ibid.*, p. 96.

10. *Ibid.*, pp. 88, 95, 101.

11. *Ibid.*, p. 353.

12. Michel Foucault, *The Birth of Biopolitics* (Basingstoke: Palgrave, 2008), p. 10.

13. Foucault, *Security, Territory, Population*, *op. cit.*, p. 45.

14. *Ibid.*, p. 108.

15. Michel Foucault, *Society Must be Defended* (Harmondsworth: Penguin, 2004), p. 28.

the macropower of the state institutions as a general mechanism of overall domination.[16] This provides us with a much more relational approach to the state that understands it as inseparable from its practices or ways of governing.[17]

If this is a general definition of governmentality, it soon becomes clear that the liberal element of rule—governing through the idea of free conduct, self-awareness and self-limitation—is a predominant feature that marks it out from other types of power or disciplinary rule. If the axis of sovereignty–discipline–government is to make any sense, then governmentality has to be understood as this increasingly liberal form of power that is engaged with, but distinct from, more centralised, territorialised and coercive forms of power. This liberalism defines a problem-space of government, its appropriate forms of regulation and its self-imposed limits. It looks to the private sphere as a way to disguise the imposition of "market discipline" as somehow an exercise in freedom. Liberal discourse presents this realm as based on the rational conduct of individuals free from state interference. However, this freedom and liberty is clearly a construction that is reinforced through a particular set of social practices and a normative discourse. Even Hayek admits that freedom is a cultural conception of something that has evolved over time, establishing a set of rules with their disciplinary effects.[18] As Dean says: "in order to act freely, the subject must first be shaped, guided and moulded into one capable of responsibly exercising that freedom".[19] This can be seen most clearly if we turn to neoliberalism and some of the work of the "Anglo-Foucauldians" like Dean, Rose and Burchell,[20] as well as Foucault's own comments on neoliberal forms of governmentality in the recently published *The Birth of Biopolitics*.

An important question to address is what the "neo" adds to liberalism. It certainly raises a question as to the naturalness and purity of liberalism if we have to distinguish between types of liberalism. As noted, pure liberalism is only an ideal type. Neoliberalism distinguishes itself precisely because of social and historical context. This context is provided by the unravelling of the post-war institutional settlement. The neoliberal discourse thus problematises the national solutions of the post-war states and argues the need to move away from centralised government activity, the welfare state and Keynesian forms of intervention. As Dean notes:

> the neo-liberal critiques of the welfare state sought to redeploy the "free subject" as a technical instrument in the achievement of governmental purposes and objectives. Contemporary liberal rule rediscovers freedom as a technical modality ... The notion of freedom and the free conduct of individuals once again becomes the principle by which government is to be rationalised and reformed.[21]

16. Michel Foucault, "Truth and Power", in James D. Faubion (ed.), trans. Robert Horley and others, *Power: The Essential Works of Foucault 1954–1984, Vol. 3* (New York: The New Press, 2001), p. 123.
17. Foucault, *Security, Territory, Population, op. cit.*, p. 277.
18. Mitchell Dean, *Governmentality: Power and Rule in Modern Society* (London: Sage, 1999), p. 157.
19. *Ibid.*, p. 165.
20. Dean, *op. cit.*; Nikolas Rose, "Governing 'Advanced' Liberal Democracies", in Miller and Rose, *Governing the Present, op. cit.*, pp. 199–218; Graham Burchell, "Liberal Government and Techniques of the Self", in Andrew Barry, Thomas Osborne and Nikolas Rose (eds.), *Foucault and Political Reason* (London: UCL Press, 1996), pp. 19–36.
21. *Ibid.*, p. 155.

This type of governmentality is all too familiar in countries such as Britain under the New Labour government. Neoliberalism engages in a process of "destatification" by introducing the norms and values of the market to other areas of social life through the promotion of competition, initiative and risk taking. As Foucault says: "The society regulated by reference to the market that the neo-liberals are thinking about is ... a society subject to the dynamic of competition ... an enterprise society."[22] Instead of direct governance, the state steps back and encourages people to become more active, enterprising and responsible for their own decisions. Burchell calls this a new form of "responsibilisation" where the governed are encouraged, freely and rationally, to conduct themselves in new ways.[23] The subjects of government are given new obligations and duties. People are appealed to as citizens or consumers who are "free" to take responsibility for their own life choices but who are expected to follow competitive rules of conduct with the logic of enterprise applied to their individual acts. In this way neoliberalism defines the boundaries of individuals and institutions and legitimates the private sphere through the belief that it is free from the interference of the state. In reality, this is achieved through neoliberalism defining positive tasks for government, constructing the legal, institutional and cultural conditions for an artificial competitive game of entrepreneurial conduct which can be applied to almost all areas of our social lives and which, as Foucault notes,[24] is guaranteed by the state. This is extended through the discourse of globalisation which provides a further spatial dimension to this process. The idea of globalisation also helps to set out what is considered inside or outside the state, what is public or private, what is within the competence of the state, what can be managed and how. Dean writes that "The distinguishing feature of reflexive government is that the point of the reform of the institutions and mechanism of government is to secure them in the face of processes that are deemed beyond governmental control."[25] In this case, it is neoliberalism and globalisation that do the "deeming". These two discourses work together to justify a deliberate set of policies by suggesting that there is no alternative but to follow the flows of capital and the logic of the free market.[26]

Governmentality and IR

This, then, sets out the social and historical context of governmentality, but what of its global nature? We have seen how the *discourse* of globalisation acts to reinforce neoliberal concerns with the free market and the regulation of human conduct and expectations. But if globalisation refers to a process that has swept across the world, does this then mean that governmentality is also something that is global in scope? The next sections seek to raise two separate issues here. This section addresses the question of whether governmentality, as a set of liberal techniques, really does apply to all parts of the globe. Following that we will look at

22. Foucault, *The Birth of Biopolitics*, *op. cit.*, p. 147.
23. Burchell, *op. cit.*, p. 29.
24. Foucault, *The Birth of Biopolitics*, *op. cit.*, p. 173.
25. Dean, *op. cit.*, p. 179.
26. For more on this see Jonathan Joseph, "Globalization and Governmentality", *International Politics*, Vol. 43, No. 3 (2006), pp. 402–418.

whether there is such a thing as a global governmentality—that is to say, not just governmentality operating in different parts of the world, but governmentality regulating the whole globe.

There are different areas where the governmentality approach can be applied to local cases. We will look briefly at security and then at the interventions by intergovernmental organisations (IGOs) like the World Bank and IMF. Certainly, the study of the operation of private security companies in different parts of the world would appear to be a prime candidate for the governmentality label. Here we have a case of privatised security operating at a distance from both the local state and foreign powers which no longer hold, or care to hold, a monopoly over the legitimate use of force and so, consequently, security spans both public–private and local–global divides.[27] With multiple actors involved, the provision of security becomes a competitive game governed by market rules. This is, of course, a game contrived by neoliberalism rather than by pure free-market rules. The result, as Leander and van Munster explain, is security governance "taking place through a set of (quasi-)markets imbued with entrepreneurial values and inspired by a hands-off approach to governance".[28] This in turn works to "depoliticise" security and to frame it, like other forms of governmentality, in a technocratic way. Looking at this through the case study of Darfur, Leander and van Munster write that:

> Within the scheme of neo-liberal governmentality the regulation of actors takes place through the employment of private sector technologies of performance such as benchmarking, best practice schemes, codes of conduct, performance indicators and auditing. In line with the view that governing through (quasi-)markets is the most effective way of dealing with problems, the purpose of these technologies is to push control out of the allegedly unaccountable and non-transparent bureaucratic sphere towards the constant scrutinizing gaze of consumers and other stakeholders such as NGOs and other humanitarian organizations.[29]

Thus far, the idea of governmentality works well in describing the *provision* of security. But this is precisely the problem in so far as the neoliberal discourse of security provision is being imposed by security actors in places quite different from those where these discourses and practices first emerge. If neoliberal practices were already contrived in their way of operating, their imposition on very different parts of the globe takes their artificiality to a new level. And in contrast to neoliberalism in advanced liberal societies, the outcome is usually quite different, often disastrously so. The proliferation of private security companies in Africa occurs not because these countries can easily be governmentalised, but usually because of the failings of public provision of security, most notably the lack of a strong and effective state capable of either directly providing security or effectively devolving its provision to others. The absence of these conditions means

27. Rita Abrahamsen and Michael C. Williams, "Introduction: The Privatisation and Globalisation of Security in Africa", *International Relations*, Vol. 21, No. 2 (2007), p. 132.

28. Anna Leander and Rens van Munster, "Private Security Contractors in the Debate about Darfur: Reflecting and Reinforcing Neo-liberal Governmentality", *International Relations*, Vol. 21, No. 2 (2007), p. 202.

29. *Ibid.*, p. 209.

that governmentality can be imposed, but it cannot develop deep roots and thus fails in its immediate aims.[30] Yet the theorisation of this usually attempts to fit security into the governmentality box, talking of how security takes a networked form, different from the old hierarchies of power, collapsing old spatio-temporal boundaries, using new information and communications technologies, set within a new risk mentality and responsible to a set of stakeholders.[31] Of course, a lot of this is true of advanced liberal societies, but little attempt has been made to show how difficult it is for such things to work outside of these societies. Little mention has been made of how the dynamics of the international society responsible for imposing governmentality are often quite the opposite of this—hierarchical, coercive and directly disciplinary. If we still wish to use the term governmentality to describe these processes, then clearly these are examples of the imposition of governmentality on societies where the social conditions are quite different from those where these techniques first emerged. This then requires us to adopt a social ontology that goes deeper than just examining techniques of governmentality and to look at the social conditions of possibility that either allow types of governmentality to develop or which can lead to more serious social problems. In contrast to most of the governmentality literature, Abrahamsen and Williams mention how

> The colonial legacy, combined with economic and political factors, have [sic] made the production of a "citizen identity" in many African countries highly problematic, and this lack of social cohesion is arguably a source of many of the continent's security problems. The privatization and globalization of security can potentially exacerbate this situation.[32]

While neoliberal forms of governmentality in the advanced liberal societies may not necessarily be desirable, we can at least see how they can operate. Outside of these advanced liberal societies where this type of governmentality has emerged, it is difficult to imagine the same techniques working effectively. When the social conditions for neoliberal governmentality are not present it is difficult to imagine governance taking place from a distance through the exercise of freedom. While the provision side may perhaps be described in these terms, the actual practices cannot be. Then our options are either to describe this process

30. For example, Abrahamsen and Williams look at private security in Kenya arguing that: "Despite its size and significance, the private security sector in Kenya is entirely unregulated and little or no attention has been paid to its role and functions. There is no specific legislation or regulation pertaining to private security companies, and no oversight or monitoring of their practices, services, and training. No special license is needed to open a security company, and it is a common complaint in the sector that it is as easy to start a security company as it is to open an ice cream kiosk. Moreover, the vast majority of security companies are not registered at all. Accordingly, the quality of companies and their services vary considerably, and there are concerns that private security companies (like the police) may, or have already, become a source of insecurity. There are frequent accounts of security guards colluding with criminal individuals and gangs in robbing their clients, although the absence of statistics makes the extent of this impossible to establish. What remains the case is that the absence of any government regulation and oversight provides no provisions for imposing sanctions, penalties or closing companies that engage in unlawful or unprofessional activities" (Rita Abrahamsen and Michael C. Williams, "Security Sector Reform: Bringing the Private In", *Conflict, Security & Development*, Vol. 6, No. 1 (2006), p. 15).

31. See Benoît Dupont, "Security in the Age of Networks", *Policing & Society*, Vol. 14, No. 1 (2004), pp. 76–91.

32. Abrahamsen and Williams, "Security Sector Reform", *op. cit.*, p. 19.

as "failed governmentality" or to explain how interventions that look as if they are based on governmentality actually revert to more coercive forms of power in regulating populations. This is precisely the point at which governmentality needs to be understood in relation to disciplinary power and the kinds of techniques, institutions and apparatuses described by Foucault in *Discipline and Punish* rather than in his governmentality lectures. While social theorists might talk up the idea of neoliberal governmentality working successfully in different parts of the world, social reality is always a harsher judge of such exercises.

The same is the case in relation to interventions by IGOs such as the World Bank and IMF. Clearly things have changed since the overtly coercive policies of structural adjustment. Instead, these institutions link support to practices of "good governance" and "capacity building" that recognise the enabling role of states in creating the best conditions for markets to function. This change of approach is evident in the World Bank's 2002 *World Development Report* which suggests that a "strong and capable state is necessary to support markets".[33] Another way this is achieved is by promoting the idea of local ownership of development projects and by trying to engage civil society and local groups. In discussing the way in which NGOs contribute to new forms of governmentality, Sending and Neumann examine how these organisations "are constituted as self-associating units through 'technologies of agency' whose political significance resides both in their capacity to convey and mobilize the preferences and concerns of individuals' and communities, and in their capacity to carry out regulatory functions".[34] Critics such as Chandler, Cammack and Kiely[35] have noted that what really happens here is that Northern-dominated institutions dictate what counts as good governance while non-Northern states are forced to take responsibility for implementing these policies. Promoting the ideas of transparency and anti-corruption allows for a depoliticised, technological approach which, along with appeals to the rule of law, can be used to blame local practices and actors if things go wrong. The consequence, according to David Chandler, is to integrate states into networks of external regulation, while also denying ultimate responsibility for the relationship obfuscating imperialist power by making the exercise of power appear as empowering rather than dominating.[36] This would seem to fit perfectly with descriptions of governmentality as the setting of boundaries for what can or cannot be done, while responsibilising local agents to "freely" do the right thing. In the words of Fine: "Education, good governance, policy ownership, and democracy are all about doing what the WB/IMF would do but also appearing to do it by yourself and willingly."[37]

33. World Bank, *World Development Report: Building Institutions for Markets* (New York: Oxford University Press, 2002), p. 36.

34. Ole Jacob Sending and Ivor B. Neumann, "Governance to Governmentality: Analyzing NGOs, States, and Power", *International Studies Quarterly*, Vol. 50, No. 3 (2006), p. 658.

35. David Chandler, *Empire in Denial* (London: Pluto, 2006); Paul Cammack, "What the World Bank Means by Poverty Reduction and Why it Matters", *New Political Economy*, Vol. 9, No. 2 (2004), pp. 189–212; Ray Kiely, "Poverty Reduction through Liberalisation? Neoliberalism and the Myth of Global Convergence", *Review of International Studies*, Vol. 33, No. 3 (2007), pp. 415–434.

36. Chandler, *Empire in Denial*, *op. cit.*, p. 77.

37. Ben Fine, "Neither the Washington nor the Post-Washington Consensus: An Introduction", in Ben Fine, Costas Lapavitsas and Jonathan Pincus (eds.), *Development Policy in the Twenty First Century: Beyond the Post-Washington Consensus* (London: Routledge, 2003), p. 12.

Of course, there is also the small matter of whether these interventions actually work. In the 1990s the World Bank recognised that there were problems with its structural adjustment programmes by attempting to bring the state back in. The new approach of "building institutions for markets" means emphasising good governance, the rule of law, efficient and transparent decision making, local ownership and effective intervention. But if the state and civil society are already weak, it is difficult to see how these new programmes can fare any better, especially if the ultimate goal remains the promotion of open markets. As with security policies, it is difficult to imagine how these imposed or implanted techniques of neoliberal governmentality can survive in a different context from that in which they initially developed. Yet they are imposed because institutions such as the World Bank and IMF are so bound up with the dominant neoliberal rationality that they are unable to see the world outside of this discursive framework. Here it is useful to bear in mind the nature of the word "governmentality". As Miller and Rose usefully suggest, the term contains two aspects of the governing process—one relating to "rationalities" or "programmes" of government, the other relating to "technologies" of enactment. One represents the world in a particular way; the other is a way of acting upon it.[38] International organisations operate according to the former even though the latter is often wholly inappropriate. This is clearly something that should not happen; yet it is the peculiar nature of the international as a series of different overlapping societies each with its own social and historical specificities. This means that developing countries suffer a modern version of combined and uneven development in so far as they are locked into the social conditions of their own stage of development, yet are subject to the strategies and techniques of the advanced liberal countries that dominate the activities of the major development organisations. It would take a significant stretch of the imagination to believe that in these cases such organisations succeed in promoting the health, wealth and well-being of populations through advanced liberal techniques of governance from a distance through the freedom and autonomy of responsibilised individual actors.

Global Governmentality?

The matter of whether these interventions by IGOs really work could perhaps be posed differently. Instead of thinking that the aim of institutions such as the World Bank and IMF is to improve the conditions of the local population, perhaps there is some other, more global motive. This broader aim would be to secure open markets across the globe. Of course, the neoliberal view is that liberal markets and poverty reduction go together; as the IMF suggests: "Countries that align themselves with the forces of globalization and embrace the reforms needed to do so, liberalizing markets and pursuing disciplined macroeconomic policies, are likely to put themselves on a path of convergence with advanced economies."[39] This argument has been questioned by a number of writers; Kiely, for example, notes that a belief in the importance of openness to global markets seriously underestimates the ability of developing countries to break into new

38. Miller and Rose, *Governing the Present*, *op. cit.*, p. 15.

39. International Monetary Fund, *World Economic Outlook* (Washington, DC: International Monetary Fund, 1997), p. 72.

export markets, suggesting that in fact these policies increase uneven development by giving competitive advantage to already developed countries.[40] But because the World Bank and IMF are wedded to the view that openness to the global market is the solution to all problems of development, then if, for some reason, such pro-market policies do not result in economic growth and poverty reduction, this must be considered something to do with the country itself, its own internal practices, lack of democracy, lack of empowerment of women and local groups, lack of transparency and over-reliance on the wrong type of regulation and state intervention.

The World Bank understands this through the idea of good governance. It suggests that

> Good governance includes the creation, protection, and enforcement of property rights, without which the scope for market transactions is limited. It includes the provision of a regulatory regime that works with the market to promote competition. And it includes the provision of sound macroeconomic policies that create a stable environment for market activity. Good governance also means the absence of corruption, which can subvert the goals of policy and undermine the legitimacy of the public institutions that support markets.[41]

Defining good governance in this way allows for a normalising discourse that sets standards by which to judge the achievement of certain domestic goals and which can be used to blame countries when these standards are not seen to have been achieved. These norms are not imposed but are applied using a complex process of assessment of compliance. Indeed, as Cammack notes, an organisation like the World Bank promotes ownership because it

> recognises that it lacks the means to enforce the strategy itself, and because the legitimation of its project *vis-à-vis* citizens around the world depends upon its adoption by national governments, which remain indispensable intermediaries in the project. But at the same time it proposes that governments should maintain a policy matrix for external inspection at any time.[42]

This notion of external inspection provides a good way of understanding governance from a distance and how states are subjected to what Mark Duffield calls "metropolitan monitoring, intervention and regulation".[43] We can look at the way IGOs compile data and indexes and use a range of benchmarks and performance indicators to assess compliance with certain rules, norms and performance targets. Various examples of these include the World Bank's World Development Indicators and Global Development Finance databases, the Millennium Development Goals Indicators, the World Economic Forum's Global Competitiveness

40. Kiely, "Poverty Reduction through Liberalisation?", *op. cit.*, p. 434.

41. World Bank Development Report, *Building Institutions for Markets* (New York: Oxford University Press, 2002), p. 99.

42. Cammack, *op. cit.*, p. 204.

43. Mark Duffield, "Social Reconstruction and the Radicalization of Development: Aid as a Relation of Global Liberal Governance", *Development and Change*, Vol. 44, No. 5 (2002), p. 1066.

Report and the OECD's Main Economic Indicators. The guiding criteria are economic ones, an example of what Foucault means when he says that governmentality takes political economy as its method of intervention. In discussing neoliberalism he writes: "The market economy does not take something away from government. Rather, it indicates, it constitutes the general index in which one must place the rule for defining all governmental action."[44] A number of governmentality theorists have done interesting work on this issue by applying it to the way government action is defined (or appraised) by international organisations. Jacqueline Best sums up this approach in arguing that a governmentality approach "provides us with some of the tools necessary to understand the ways in which these political economic imperatives have been internationalized and institutionalized in recent years—through the non-juridical logic of international standards, the calculating metric of transparency and the entrepreneurial ethic of self-responsibility".[45] The issue of transparency is particularly interesting as a way of disciplining states and economies, and IGOs publish a range of indicators to scrutinise whether different countries have managed to meet satisfactory performance targets or to compare how well countries have managed in relation to one another. In other words, neoliberal governmentality constitutes states on the basis of global standards of conduct and competitiveness rather than seeing them as socio-political entities.[46]

What is interesting about these arguments from the point of view of a global governmentality approach is that they are now focusing on states rather than populations as the target entities. How consciously the theorists do this is open to discussion and debate. Fougner is clearest in stating that governmentality is not only about how states and governments act on populations but also how global institutions act on states:

> While much governmentality research has focused on how neoliberalism has come to inform multiple practices on the part of state authorities, the argument here is that states are themselves increasingly subjected to a form of neoliberal governance in the contemporary world political economy—in the sense that they are constituted and acted upon as subjects with a rationality derived from arranged forms of entrepreneurial and competitive behaviour.[47]

44. Foucault, *Birth of Biopolitics*, *op. cit.*, p. 121.

45. Jacqueline Best, "Why the Economy is Often the Exception to Politics as Usual", *Theory, Culture & Society*, Vol. 24, No. 4 (2007), p. 102.

46. Tore Fougner, "Corporate Power in World Politics: The Case of the World Economic Forum", *Journal of International Trade and Diplomacy*, Vol. 2, No. 2 (2008), p. 118.

47. Tore Fougner, "Neoliberal Governance of States: The Role of Competitiveness Indexing and Country Benchmarking", *Millennium*, Vol. 37, No. 2 (2008), p. 308. Fougner goes on to talk of this in relation to benchmarking: "First, given its provision of 'an external frame of reference explicitly linked to concerns about competitiveness', benchmarking constitutes states as competitive entities driven not by internal socio-political processes, but rather by external or global standards of conduct. Second, given the importance ascribed to quantitative measures and comparisons of performance, benchmarking constitutes states as calculative agencies, or entities with a capacity to calculate and rank alternative courses of action. Third, given the overriding concern with implementing 'best practice', benchmarking constitutes states as technocratic agencies acting in accordance with expert determination of what works best. Fourth, given the centrality of change and continuous improvement—as a consequence of how standards or 'best practices' undergo continuous change—

This is clearly quite a different issue from that of how governmentality works on local populations, and it returns us to our starting point: how we define governmentality and what it is that governmentality is referring to. For Fougner we have some sort of governmentality once removed in so far as the issue is not really that of the regulation of populations; indeed, it may not even matter that liberal techniques of governmentality do not work on populations in a non-liberal context if global governmentality can successfully regulate the behaviour of states. Zanotti has a similar global focus in examining how: "'Good governance' constructs both states and the international arena as governmentalized space ... good governance doctrines promote institutional arrangements that foster the reorganization of an array of local practices."[48] However, she emphasises the importance of population through a focus on Foucault's concept of biopower as the concern for the population's "common good". With global governance, therefore, this concern with the life of the population is considered to have been taken into the international arena. This is then examined through things such as the United Nations' Millennium Development Goals. Here there is perhaps less focus on the regulation of states and the question is raised as to whether it is really states or their populations who are being targeted. Merlingen, in his study of IGOs, perhaps looks both at states and their populations, but again we find a mixing of the concepts of governmentality and biopower:

> Whatever the form and target of IGO interventions in a country—its social body and the institutions enframing it—such interventions are always allied to certain kinds of knowledge about the political, economic and social characteristics of the place and its people. Biopower brings "life and its mechanisms into the realm of explicit calculations and [makes] power-knowledge an agent of transformation of human life" (Foucault, 1998: 143).[49]

The introduction of the notion of biopower brings another difficult issue to the fore since biopower (or more precisely, biopolitics[50]) might clearly be distinguished from sovereign power in having population rather than territory as its object; but biopower is not the same thing as governmentality. Indeed, biopower may be said to be refined by liberal techniques of governmentality, but may also include more coercive or disciplinary forms of power. Therefore, using the terms biopower and governmentality interchangeably runs the risk of widening the focus too much so that the specificity of a liberal notion of governmentality gives way to an all-embracing idea of biopower. Foucault himself felt the need to move away from a general discussion of biopolitics to explain the specific ways in which liberalism works to rationalise the exercise of government.[51]

benchmarking constitutes states as transformative agencies, or entities engaged in a never-ending process of reinventing themselves" (*ibid.*, p. 319).

48. Laura Zanotti, "Governmentalizing the Post-Cold War International Regime: The UN Debate on Democratization and Good Governance", *Alternatives*, Vol. 30, No. 4 (2005), p. 479.

49. Michael Merlingen, "Governmentality: Towards a Foucauldian Framework for the Study of IGOs", *Cooperation and Conflict*, Vol. 38, No. 4 (2003), p. 368.

50. Biopower includes the notions of biopolitics—concerned with populations—and anatomopolitics—concerned with individuals.

51. Foucault, *Birth of Biopolitics*, *op. cit.*, pp. 317–318.

There is no doubt that the idea of biopower can be useful in bringing into the study of IR a concern with "the problems posed to governmental practice by phenomena characteristic of a set of living beings forming a population".[52] This can even be linked to the actions of international organisations. But can it help describe the precise ways in which "States became the subject of international scrutiny and reformation efforts aimed at making them function as 'governments' instead of as uncontrolled 'sovereigns' ... and at making them the visible and predictable actors of a normalized international arena"?[53] Zanotti thinks it can, but in doing so there is a danger of losing the specificity of governmentality. When she writes that governmentality "emerges as a multifaceted and universally valid technique of rule, a knowledge/power formation that opens multiple spaces of visibility at the national and international level"[54] our first thought should surely be to recognise the gap between the rationality of governmentality and its actual realisation. For in practice, neoliberal governmentality *cannot* be a universally valid technique, for the underlying causal reasons already mentioned. It fails in many parts of the world precisely because it is unable to operate effectively outside of the social conditions of advanced liberal capitalism. The "good governance" approach of international organisations makes the *claim* to being a "universally valid technique of rule", but actual practice shows this not to be so. It is the job of the critical theorist to expose the gap between myth and reality.[55]

Conclusion

We can now start to answer a few questions. "Governmentality of what?" should always mean governmentality of populations. While Foucault's work on governmentality does not always provide straightforward explanations of exactly what it is, we can at least work out a few basic guidelines. Foucault is clear that governmentality emerges in modern societies when populations become the object of government and are addressed in terms of their health, wealth and well-being. This, we could say, represents the birth of biopolitics. These changes relate to the development of capitalism and a new set of concerns relating to populations as citizens and workers. Political economy emerges as the means by which

52. *Ibid.*, p. 317.

53. Zanotti, *op. cit.*, p. 480.

54. *Ibid.*

55. That many Foucauldians do not develop a critical approach to the actual expressions of governmentality can be attributed to their refusal to engage with deeper underlying social relations. As Zanotti says of her approach: "Instead of asking under which conditions and through what kind of interventions democratization can best be achieved, it uses the tools developed by Foucaultian studies on government to explore the conditions of emergence of good governance as the UN political rational [sic], the mechanisms of government it promotes, and the political effects it produces" (*ibid.*, p. 462). This is a good example of the tendency to give priority to an analysis of the rationality of governmentality rather than the social conditions within which it operates. Larner and Walters are even more explicit: "What we have called global governmentality entails a move of 'bracketing' the world of underlying forces and causes, and instead examining the different ways in which the real has been inscribed in thought" (Wendy Larner and William Walters, "Global Governmentality: Governing International Spaces", in Wendy Larner and William Walters (eds.), *Global Governmentality: Governing International Spaces* (London and New York: Routledge, 2005), p. 16). For a scientific realist account of underlying causal relations see Jonathan Joseph, "Foucault and Reality", *Capital & Class*, Vol. 82, No. 2 (2004), pp. 141–163.

these questions are to be addressed. Governmentality increasingly focuses on dealing with social questions through the promotion of rational self-conduct. In doing so, governance is always reflexively aware that it may be "governing too much"; governmentality therefore develops as a means to devolve powers and promote frugality of government. To deal with the limits of governmentality, Foucault places it in an axis of sovereignty–discipline–government. This means that liberal techniques of government, which seek to work from a distance through the promotion of the free conduct of individuals, can always give way to more direct forms of disciplinary power. It also keeps the state in place as a crucially important site of power, but examines the state not just through the lens of sovereignty but also in relation to disciplinary techniques and the governmentalisation of state power. The latter can be seen in the way that states have devolved a wide range of duties and responsibilities, while maintaining steering capacity. Indeed, for Foucault, the process of governmentalisation is what allows the state to continue to play a dominant role.[56] This relationship should be stressed as working in both directions. First, a governmentality approach shows how state power adapts and evolves, but equally a study of the state, as a complex institutional ensemble, helps to explain how and why governmentality works. Or in some cases raised here, a study of the weaknesses of state institutions reveals why governmentality is not effective in certain places.

Governmentality is defined by its social and historical context and has to be seen not as a thing but as a process. Thus contemporary forms of governmentality have to be seen in relation to the emergence of neoliberalism and the response to the unravelling of the post-war institutional settlement. This can be seen in relation to both national forms of economic regulation and state intervention, and the international regimes of economic and financial stability associated with the Bretton Woods system. While it is important to look for regional variations, clearly the dominant form of governmentality is this neoliberal version. Among its essential features is a further questioning of the limits of state power and a focus on the market through the introduction of rules of competition and the construction of an entrepreneurial model of conduct. While neoliberalism promotes the freedom of individual conduct, this conduct is "responsibilised" and urged to be reflexive about its own behaviour. We have seen how national governments have sought to introduce policies through a promotion of strategies and techniques of competition, risk taking, insurance, benchmarking and best practice. This is combined with more sophisticated techniques of data gathering and surveillance in order to regulate populations from a distance.

All this can be seen in the advanced liberal countries as well as in the development of regional institutions such as the European Union. However, these are the rich developed countries where a particular set of social conditions has given rise to these specific techniques. The problem with a governmentality approach is trying to work out how these techniques apply to less developed countries which clearly do not have the sorts of social relations necessary for sustaining such techniques of governance. It would be mistaken to believe that the theory of governmentality can be transposed to these countries in order to explain the regulation of their populations. Nevertheless, it would be wrong to abandon the idea of governmentality, given the particular way that Northern states and

56. Foucault, *Security, Territory, Population*, *op. cit.*, p. 109.

international organisations attempt to intervene. This leaves a situation in which governmentality appears not to work in certain parts of the world, yet where international organisations seek to intervene precisely on this basis. Clearly the task that flows as a consequence of this—something not achieved by most IR approaches to governmentality—is to explain governmentality through its failure and to point to the way that techniques developed in one part of the world have been imposed on societies with quite different social conditions as a form of the exercise of power by Northern-dominated institutions. In many cases, therefore, the focus for IR scholars should be on the obstacles to the governmentalisation of populations, states and societies—something that requires an ontological shift from explaining "how" technologies may work to the deeper social relations that explain the "why" of governmentality.

Explaining why these techniques are still imposed is in part a matter of exposing dogma and social conditioning. International organisations are as much a reflection of a particular rationality of governance as they are instigators of one. Or at least the particular bureaucrats, officials and policy makers who populate these institutions are themselves subjects whose understanding has been constructed within a particular epistemic field which makes them see the world in a particular way even if this is wholly inappropriate to problem solving in less developed countries. But if at one level of analysis we can explain these interventions in terms of discursive conditioning, at another level these interventions are more deliberate and have a different target. If the idea of global governmentality is to have any sort of meaning then it should be redefined as techniques aimed at regulating the behaviour of states and governments. We have seen how this takes different forms—benchmarking and targets, practices like good government and transparency and openness to the discipline of global markets. And as the emerging IR literature shows, it is carried out by IGOs and other organisations.

This returns us to our initial question: governmentality of what? For surely we have just insisted that governmentality is tied to the management of populations? How, then, can we accept arguments by Merlingen, Fougner, Zanotti and others that global governmentality targets states? There are significant problems with making this shift in the level of analysis, not least the anthropomorphic one of treating states as people.[57] The solution, if we are to maintain that at one level of analysis it is states rather than people and groups who are subjected to governmentality, is to argue that the regulation of states takes place *through* the targeting of populations. The fact that governmentality is usually unsuccessful at regulating populations does not matter if this can be used as a means to manage states. The uneven nature of the international means that techniques developed in one part of the world may unsuccessfully be applied in a different part of the world. Unevenness also means that states coexist in hierarchical power relations. We are now in a position to draw our rather paradoxical conclusion: global governmentality is mostly about the unsuccessful regulation of populations and that it is precisely by virtue of this that the successful regulation of states can occur.

57. Enthusiastically embraced by Alexander Wendt, "The State as Person in International Theory", *Review of International Studies*, Vol. 30, No. 2 (2004), pp. 289–316.

Foucault's Concept of Power and the Global Discourse of Human Rights

IVAN MANOKHA

This article carries out a Foucauldian analysis of the global discourse of human rights. In the spirit of Foucault's genealogy it identifies a historical discontinuity in the development of human rights between the Cold War period, when human rights were a heavily contested concept, and the late modern world in which human rights are becoming a form of global standard whose validity is less and less questioned. Using Foucault's understanding of the way power is exercised within discursive structures and knowledge/power configurations, the article argues that in the late modern world human rights constitute a global norm with reference to which agents are evaluated and increasingly evaluate themselves. Power may be exercised over those forms of agency that do not conform to this norm by other agents, and at the same time, we witness more and more situations in which there is no such coercion and yet agents find it necessary to alter their behaviour and declare their adherence to human rights. The article also seeks to extend Foucault's framework in order to explain the emergence of the global human rights discourse. It thereby complements a Foucauldian analysis with a political economy approach which, it is argued, helps us enhance the Foucauldian framework which suffers from the separation of discourses and politics from the economy.

Introduction

The aim of this paper is to carry out a Foucauldian analysis of a global discourse of human rights that has developed in the late modern global political economy (GPE). Foucault's notion of discourse allows us to bring together linguistic and extra-linguistic elements of human rights, that is, the rhetoric of human rights on the part of numerous forms of agency (state leaders, civil society activists, business executives, academics, journalists, etc.) and different policies that are advocated or adopted by them in the name of human rights.

Since the end of the Cold War, which has been perceived largely as a victory of freedom over tyranny, a commitment to further combat oppressive regimes, to spread democratic values and institutions, and to promote individual rights—mostly political and civil rights—has been expressed by a multiplicity of different state and non-state actors at international, regional, national and local levels. A whole range of policies and strategies of human rights advocacy and protection—ranging from campaigns and rallies, to making the respect for human rights a pre-condition for economic aid, to sanctions and even the use of military force—have been developed and implemented. The development of this discourse of human rights over the last two decades, as will be argued below, has resulted in rendering human rights a form of global standard or norm with reference to which different agents may be evaluated and disciplined, and with respect to which they increasingly evaluate themselves, and change their behaviour in order to conform (at least officially) to the standard.

The examination of situations in which dominant discursive structures provide a set of norms with reference to which agents may exercise power over other agents, as well as over themselves, to ensure conformity has been at the core of Foucault's historiography, particularly in his later works *Discipline and Punish* and *The History of Sexuality.* The traditional analyses of power in political science and its subfields such as International Relations (IR) have focused on its negative and behavioural aspects, that is, on power as observable repression, coercion or negation. On this view, power is the ability of an actor A to make an actor B do things (that are, as a rule, in A's interests) which B would not otherwise have done. Power on this reductionist view is a property of an agency; it is *possessed* and used by A to obtain the desired behaviour change in B. One of the key contributions of Foucault's project has been to introduce a different conception of power, which is structural and which is not reducible to the properties, intentions or characteristics of agents. This form of power results from a constellation of discursive structures, (scientific) knowledge and practices that accompany them which create a set of rules and standards, with reference to which agents may *exercise* power over other agents, as well as over themselves. This conception of power has two principal characteristics: (1) power is not possessed by agents; when they use power they *mediate* the dominant view of what constitutes normality or deviance. To put it differently, here A and B are not fixed, they constitute forms of agency that are *substitutable*; what matters is the discursive structure that sets the standards of accepted or expected behaviour; and (2) power is positive, it "produces" behaviour that is in conformity with the dominant standard of normality or acceptability. This means that power may be exercised not only over others but also over oneself, a situation in which the subject transforms himself or herself into an object of power and adopts forms of behaviour that are expected by the prevailing discourse and truth configurations. To return to the example of A and B, A may be totally absent, and yet, despite the fact that there is no exercise of power by A, B will nevertheless alter its conduct.

It will be shown below that these two elements of power characterise the "games of truth", to use another of Foucault's term, that take place within the global discourse of human rights. First, in line with Foucault's genealogical method, we will see that with the end of the Cold War a historical discontinuity may be identified with respect to human rights. During the Cold War, human rights—particular civil and political liberties of individuals—was a heavily contested concept. It was challenged, on the one hand, by socialist states on the grounds that it represented a bourgeois version of freedom and justice, and, on the other, by non-Western developing political economies as a Western concept unfit or inapplicable in other cultures characterised by the primacy of communal interests and values. With the end of the Cold War, the concept of human rights, understood predominantly as negative rights of individuals, is less and less contested, and is increasingly acknowledged as one of the priorities of governments around the world. In addition to this, human rights are increasingly promoted by business enterprises, which was not the case previously. To put it differently, human rights may be said to have become a global standard or a global norm with reference to which different forms of agency are evaluated and may be disciplined in various ways by other agents.

Thus, with respect to the first characteristic of Foucault's notion of power we will see that the subjects and objects that power exercises with reference to human rights are substitutable: it may be states disciplining other states or business enterprises; it may also be business enterprises engaging in policies of human rights promotion in states where they are not respected; it may be non-governmental organisations (NGOs) organising campaigns against states or business enterprises to enforce respect for human rights. Concerning the second aspect of Foucault's notion of power, it will be shown that today we increasingly witness changes in the behaviour of states or corporations which proclaim their adherence to human rights norms and adopt policies of their promotion without any coercion. That is to say, today agents tend to find it necessary to declare their commitment to the respect of human rights, to conform to the established standard of what constitutes an accepted polity, and increasingly an accepted business activity. This, of course, is not to say that non-Western states have suddenly started respecting human rights, and that corporations not only avoid engaging in practices that endanger human rights but in fact have become their champion. It is only to identify a qualitative change or a discontinuity with respect to what was the case during the Cold War, which consists in (a) a gradual disappearance of a critique of the concept of human rights and (b) its development into a global norm with reference to which agents are evaluated and increasingly evaluate themselves.

Such an approach to human rights constitutes an alternative to the existing analyses, particularly of those that are inspired by Realist theory, which tend to treat human rights as a cover for the selfish interests of actors that engage in their promotion or enforcement. While it may well be the case that in some cases human rights are used only to conceal some other agendas, it does not explain why it is human rights and not some other notions that are employed. Furthermore, such analyses usually treat power that is used, most often by Western states, in the name of human rights in a reductionist manner, as their property or capability, as something that is possessed. Using Foucauldian analysis enables us to have a better insight into such uses of power; it helps us to see that it is not reducible to actors that engage in human rights enforcement, and that in fact what they do is *mediate* a global norm.

However, despite the usefulness of Foucault's categories and method, the analysis of the global discourse of human rights needs to go beyond the framework of Foucauldian categories if we wish to *explain* the emergence of this discourse. That is to say, if we wish to understand why it is that this particular configuration of knowledge and individual rights that has come to dominate the late modern GPE, and not just focus on the way it operates, we have to extend our analysis beyond Foucault's categories. Thus, it will be argued below that Foucault's historiography and his method of genealogy is useful in identifying the supersession of some discursive structures by others, in highlighting differences between them, as well as the implications of these differences for practice. However, a strictly Foucauldian framework would tend to under-privilege explanations for such changes, treating these historical breaks as a starting point, without seeking to subject to scrutiny the reasons for the emergence of new historical configurations, and their role in social totality. In his attempt to provide an alternative to Marxism and its method of historical materialism, Foucault neglects the role of economic relations, of structures of production which, as will be argued below, constitutes a regrettable omission. We will seek to complement Foucauldian analysis with an account of the development of the global discourse of human rights which examines it with reference to production relations. It will be argued below that at the heart of the global discourse of human rights reside individual rights that emerged with the rise of capitalism, that have played a causal role in its development, and that continue to play this causal role today. With the help of the concept of ideology used in the negative sense, we will see that the dominant view of what constitutes (fundamental) human rights identifies individual political and civil liberties that are not only compatible with the operation of global capitalism but in fact play an important causal role in its reproduction. It will be argued that at the current historical juncture the global discourse of human rights provides the existing order at the heart of which reside the interests of transnational capital with a Gramscian quality of moral and intellectual leadership.

Foucault's Historiography

To understand Foucault's project it is necessary to locate it in its historical context. Beginning with the late 1950s, Western Marxism, which dominated the field of critical theory and opposition in Western Europe, could no longer provide an adequate analysis of advanced industrial societies. On the one hand, the Marxist conceptualisation of politics in terms of class politics and its reduction of the political to the level of class alliances and class struggle were hardly suited to the examination of diverse forms of oppression and resistance movements that increasingly characterised Western societies. Such phenomena as discrimination against sexual minorities, the subordinate position of women, the overwhelming and omnipresent power of bureaucracy or the treatment of prisoners and the mentally ill could not be reduced to a project of the bourgeoisie and required a different conceptual framework. On the other hand, the opposition to these forms of oppression was not carried out by the proletariat either, but by a whole range of different social movements: the women's movement, the student movement, the gay liberation movements, the movement for prison reform, the anti-psychiatry movement, and so on. These shortcomings of Marxism as a method of analysis and as a theory of praxis manifested themselves most acutely during the May 1968 protests in France. During the month of May the protest movement was constituted by social groups not traditionally associated with the proletariat: the events were initiated by students, and carried on by professional and technical workers; in terms of their demands, they spoke not only against capitalism but also against bureaucracy and all forms of non-democratic social organisations. As Lefort has argued, "the protesters questioned the 'organizational viewpoint' which ordered Western society by parceling social life and labeling individuals".[1] In addition to this, there was a more general dissatisfaction with Marxism as a political project in the light of the highly oppressive Soviet state and the barbarity

1. C. Lefort, "Then and Now", *Telos*, No. 36 (1978), p. 37.

of its military interventions abroad justified as necessary to preserve socialism (in East Germany in 1953, in Hungary in 1956, and in Czechoslovakia in 1968).

The dissatisfaction with Marxism and with the method of historical materialism is at the heart of Foucault's project. His work can be described as an attempt to make intelligible modes of domination or "technologies of power" that escaped the attention of Marxism, and, in order to do so, to develop a different form of historiography and an alternative conceptual apparatus. Foucault was particularly concerned with the operation of discourses dominated by science, and their relationship with social practice and forms of oppression that result from these configurations. For Foucault, historical materialism was inadequate in that it split discourse from practice and then subordinated the former to the latter, and thereby introduced a division in critical theory between what human beings say and what they do, a division which could no longer go unchallenged.[2] At the same time, by designating itself "science", Marxism gave itself a false legitimacy that enabled Marxist theorists to place themselves above the masses as the bearers of the universal. The method of historical analysis that Foucault adopts instead is the method of "genealogy" of Friedrich Nietzsche. Nietzschean genealogy was an effort to delegitimise the present by separating it from the past. Contrary to "traditional" historical studies which examine particular events within grand explanatory systems and processes seen as linear evolutions, Nietzschean genealogy seeks to depict the present as finite, limited, even repugnant, simply by locating differences in the past. The Nietzschean historian begins with the present and goes backwards in time until a difference is located. Then he proceeds forward again, tracing the transformation and taking care to preserve the discontinuities as well as the connections in the historical line.[3] As Foucault states with reference to Nietzsche's genealogy:

> what Nietzsche has incessantly criticised ... is the form of history ... which would allow us to recognise ourselves everywhere [in history] and to give to all shifts and changes a form of reconciliation; ... To know, even in history, does not mean to "recognise", and certainly not "recognise ourselves". History may be "effective" only in so far as it succeeds in introducing the discontinuous into our own current being.[4]

Foucault's genealogy is similarly an attempt to study historical discontinuities and breaks, rather than chronologically narrate what happened in the past; its goal is "to establish and preserve the singularity of events, turning away from the spectacular in favour of the discredited, the neglected, and a whole range of phenomena which have been denied a history (e.g. reason, punishment, sexuality)".[5] As Foucault observes, the objective of genealogy is to

> to spot the uniqueness of events, outside any monotonous finality; to look for them where we would least expect them and in what goes as having no history at all—feelings, love, conscience, instincts; to capture their

2. M. Poster, *Foucault, Marxism and History: Mode of Production versus Mode of Information* (Oxford: Polity, 1984), p. 66.
3. *Ibid.*, p. 64.
4. M. Foucault, *Anthologie* (Paris: Gallimard, 2004), p. 409; my translation.
5. B. Smart, *Foucault, Marxism and Critique* (London: Routledge, 1983), pp. 75–76.

> return, in order not to trace the curve of their slow evolution, but to find different stages at which they played different roles.[6]

Thus Foucault does not focus on the evolution of the past, on the story of how historical events eventually lead into the present, but seeks to show how the past was different, strange and at times threatening.

What is important to note is that Foucault's reflections on genealogical analysis demonstrate that it is inextricably linked with a commitment to critical analysis, to the discovery of forms of oppression and power, and the manner in which individuals become the subjects and objects of technologies of power. For Foucault, knowledge is political in the sense that its conditions of existence include power relations. As he puts it with respect to his genealogy of sexuality:

> by genealogy I do not mean to provide a history of successive conceptions of desire, lust or libido, but to analyse the practices by means of which the individuals have been driven to examine themselves, to decipher themselves, to recognise and admit themselves as objects of desire ... In short, the idea was, in this genealogy, to look for ways in which individuals have been driven to exercise over themselves and over others the hermeneutics of desire.[7]

The genealogical analysis of discontinuity of knowledge/power configurations within which individuals exercise power over others and over themselves is particularly pronounced in Foucault's later works *Discipline and Punish* and *The History of Sexuality*. In these two pieces Foucault divides the histories of crime and sex into two or three main periods, with the 18th century serving as the dividing line. In the earlier periods, the focus of knowledge/power about crime and sex was the body of individuals; beginning with the 18th century, it shifts to the mind. New regimes are established in which criminals and sexual activity are scrupulously monitored. Elaborate bureaucracies are established to keep tabs on people; files are developed with an expansion of disciplines and scientific experimentation to study the most banal thoughts and actions of potential criminals and recidivists, childhood masturbators and hysterical women.[8]

Now, returning to the concept of power, Foucault observes that we may distinguish two categories of power: (1) the techniques of domination "which allow to define the behaviour of individuals, to impose certain finality or certain objectives"; and (2) the techniques of the self "which allow individuals to carry out, by themselves, a certain range of operations on their bodies, their souls, their thoughts, their conduct, and to do so in order to produce in themselves a transformation, a modification, and to attain a certain state of perfection, of happiness, or purity, of supernatural powers."[9] Taken together, these two categories form what Foucault referred to elsewhere as "governmentality" or "the encounter between the techniques of domination exercised over others, and the techniques of the self".[10] And, as already mentioned, a crucial role in the development of

6. Foucault, *op. cit.*, p. 393; my translation.
7. *Ibid.*, p. 708; my translation.
8. Poster, *op. cit.*, p. 81.
9. Foucault, *op. cit.*, p. 654; my translation.
10. *Ibid.*, p. 655; my translation.

governmentality is played by sciences. As Foucault states: "From the eighteenth century until today, 'human sciences' have reintegrated the techniques of verbalisation into a different context, making of them ... the positive instrument of the constitution of a new subject."[11]

Thus, what we may infer from the foregoing discussion is that Foucault's conception of power may be said to have two key characteristics. On the one hand, power is not reducible to agents that exercise it, it is not their property; when power is exercised it *mediates* the dominant view of what constitutes normal and acceptable agency, and forms of behaviour that are seen as deviant or unacceptable, and the form it takes is based on the prevailing conventions which determine how deviance is to be dealt with. As Foucault puts it, individuals do not possess power but constitute its effect, "the element of its articulation ... its vehicle".[12] As mentioned above, on this view the subjects and objects of power are interchangeable; they operate within a "discourse with a truth-function, discourse which passes for the truth" which "constitutes a set of structures *whose agency is at once everyone and no one*".[13]

On the other hand, Foucault describes forms of power that are not based on negation or repression, but which are positive, which "produce reality and produce domains of objects and rituals of truth".[14] It refers to situations in which power is exercised not only over others but also over oneself; it depicts how the subject transforms himself into an object of power and adopts "willingly" forms of behaviour that are expected by the prevailing discourse and truth configurations. As Foucault argues, the role of historiography is to establish

> voluntary practices by means of which individuals not only create for themselves the rules of conduct, but also endeavour to transform themselves, to modify their unique being ... and [by means of which] an individual comes to think of his own being when he recognises himself as mad, when he regards himself as ill, when he thinks of himself as a living being, working and talking, when he judges and punishes himself as criminal.[15]

In sum, what I wish to highlight for the purposes of the discussion of the global human rights discourse that will be undertaken in the next section is that Foucault provides us with tools to analyse historical configurations of power structures, historical breaks or discontinuities in their existence, as well as the manner in which agency is constituted and disciplined within them. Of particular importance to what follows are Foucault's insights into the manner in which particular discursive structures and knowledge configurations develop norms, rules and standards of accepted or normal behaviour, with respect to which agency may be evaluated and evaluates itself. Let us now see how these notions can help us understand the global discourse of human rights.

11. *Ibid.*, p. 663; my translation.
12. M. Foucault, *Power/Knowledge: Selected Interviews and Other Writings 1972–1977* (Brighton: Harvester Press, 1980), p. 98.
13. Poster, *op. cit.*, p. 78; emphasis added.
14. Smart, *op. cit.*, p. 86.
15. Foucault, *Anthologie*, *op. cit.*, pp. 709, 713–714; my translation.

A Foucauldian Analysis of the Global Discourse of Human Rights

Human rights appeared on the international scene after the end of the Second World War with the adoption of the UN Charter in 1945 and the proclamation of the Universal Declaration of Human Rights in 1948. However, right from the start the concept of human rights was challenged by the USSR and its allies as an expression of individualistic bourgeois morality. As Evans observes, the socialist states asserted that individual rights "were nothing less than a reactionary attempt to legitimate a set of ... bourgeois values that did little for the poor and excluded".[16] Indeed, an irreconcilable tension existed between the primacy given to the individual embodied in the notion of human rights and the privileging of the collective that was at the centre of socialist systems. One of the rights that posed perhaps the most difficulties was the right to private property, which was not recognised in the Soviet bloc. The USSR Constitution, for example, stated that the economic foundation of the USSR was "the abolition of private ownership of the means of production" (Article 4) and that property existed "either in the form of state property or in the form of co-operative and collective-farm property" (Article 5). The socialist camp accused the West of economic exploitation of people and advocated the international development of social and economic rights.[17] This advocacy of positive rights received a new momentum with the process of decolonisation as newly independent states expressed various reservations concerning the concept of individual rights and joined the socialist countries in the critique of the Western conception of human rights.[18] Although more often than not such critique was a strategy employed by authoritarian regimes to justify their non-democratic systems as conducive to political stability which, it was claimed, was necessary for economic development and the guarantee of economic and social rights (deemed to be a more important objective than democracy), it is nevertheless important to note that such regimes contributed to the challenge of the notion of individual political and civil rights.

In addition to this, the Cold War environment characterised by the constant competition for allies and attempts on the part of the two blocs to extend their spheres of influence meant that even Western states sacrificed human rights considerations when they came into conflict with their strategic interests.[19] It is also worth noting that the three cases of intervention that are regarded by a number of scholars as examples of humanitarian intervention that took place during the Cold War—the Indian intervention in East Pakistan in 1971, the Tanzanian intervention in Uganda in 1979, and Vietnam's intervention in Cambodia in 1978—involved only marginal references to human rights. Most of the justifications for these actions centred on self-defence and retaliation.[20] In short, human rights were a heavily contested concept.

16. T. Evans, *The Politics of Human Rights: A Global Perspective* (London: Pluto Press, 2001), p. 23.

17. A. Rosas and J. Helgensen, "Introduction", in A. Rosas and J. Helgensen (eds.), *Human Rights in a Changing East–West Perspective* (London: Pinter, 1990), p. 1; P. Hunt, *Reclaiming Social Rights: International and Comparative Perspectives* (Aldershot: Dartmouth, 1996), p. 7.

18. M. Freeman, *Human Rights: An Interdisciplinary Approach* (Cambridge: Polity Press, 2002), p. 47.

19. K. Feste, *Expanding the Frontiers: Superpower Intervention in the Cold War* (New York: Praeger, 1992), p. 1.

20. See F.K. Abiew, *The Evolution of the Doctrine and Practice of Humanitarian Intervention* (London: Kluwer, 1999); T. Franck and M. Rodley, "After Bangladesh: The Law of Humanitarian Intervention by Military Force", *American Journal of International Law*, Vol. 67 (April 1973), pp. 275–305.

With the end of the Cold War and the transition of most of the former socialist countries to capitalism, the structural incompatibility between individual rights and their socio-economic systems disappeared. The adoption of policies of privatisation and the replacement of the command economy with a market economy went hand in hand with policies of democratisation and the institution of individual rights. Thus, the Russian Constitution of 1993 proclaims various civil and political rights, including the right to private property, as the foundation of the Russian state. In addition to this, the disappearance of the Cold War "geo-political straightjacket"[21] made the sacrifice of human rights for strategic interests much less necessary. The West, which proclaimed the end of the Cold War as the victory of democracy, freedom and individual rights, invested itself with a mission to promote these ideals elsewhere around the world. Of particular significance have been the human rights policies of the United States and the United Kingdom. Thus, after the Cold War, and especially during the presidency of Bill Clinton, human rights were declared the central pillar of American foreign policy: "Advancing human rights must always be a central pillar of America's foreign policy."[22] In the case of the United Kingdom, with the coming to power of a Labour government in 1997 an "Ethical Foreign Policy" was proclaimed by Robin Cook, who stated the following: "The Labour Government will put human rights at the heart of our foreign policy and will publish an annual report on our work in promoting human rights abroad."[23]

Numerous human rights initiatives were also undertaken by different international organisations, including those whose initial functions had nothing to do with human rights. For example, the Bretton Woods institutions incorporated human rights into their agendas through the promotion of what they have referred to as "good governance". As the former managing director of the IMF put it, "a global economy needs global ethics, reflecting respect for human rights ... and the IMF and World Bank are part of the workforce to make a better globalization".[24] As Chandler observed in this respect, the IMF and the World Bank "whose mandates seem to be unrelated to human rights, have integrated these concerns and acted on them in ways unthinkable ten years ago".[25]

The United Nations, no longer paralysed by the US–Soviet ideological rift, embarked upon global human rights promotion in a multiplicity of ways. There is no space here to discuss all the UN human rights policies and programmes.

21. A. de Waal, *Famine Crimes: Politics and Disaster Relief Industry in Africa* (Oxford: James Currey, 1997).

22. W. Clinton, "Remarks in Honour of Human Rights Day", Museum of Jewish Heritage, New York, 9 December 1997, US Department of State, available: <http://www.state.gov/www/global/human_rights/971209_clinton_humrts.html>. See also *idem*, "Remarks at Human Rights Day Presentation of Eleanor Roosevelt Human Rights Award", 10 December 1998, US Department of State, available: <http://www.state.gov/www/global/human_rights/981210_whps_dayhr.html> (all URLs accessed 5 July 2009).

23. R. Cook, "Robin Cook's Speech on the Government's Ethical Foreign Policy", *Guardian Unlimited* (12 May 1997), available: <http://www.guardian.co.uk/indonesia/Story/0,2763,190889,00.html> (accessed 5 July 2009).

24. H. Köhler, "Working for a Better Globalization", Remarks at the Humanizing the Global Economy conference, Washington, DC, 28 January 2002, International Monetary Fund, available: <http://www.imf.org/external/np/speeches/2002/012802.htm> (accessed 5 July 2009).

25. D. Chandler, *From Kosovo to Kabul: Human Rights and International Intervention* (London: Pluto Press, 2002), p. 8.

To mention only a few initiatives, the United Nations organised a World Conference on Human rights in Vienna in 1993, at which the Office of the UN High Commissioner for Human Rights was created, and institutional foundations laid for UN partnership with NGOs. This was soon followed by partnership with business firms within the framework of the Global Compact launched by Kofi Annan in 2000. Another important development has been the establishment by the UN Security Council of two international tribunals for trials of crimes against humanity and genocide perpetrators in Former Yugoslavia and Rwanda, which were further innovations by the United Nations. These two ad hoc tribunals stimulated proposals to create a permanent court that would deal with gross violations of human rights and crimes against humanity. Such proposals materialised in the establishment in 2002 of the International Criminal Court within the framework of the United Nations. Furthermore, the UN Security Council authorised a number of military actions to enforce human rights. I will return to these instances of the use of force below in the discussion of the manner in which power has been exercised with reference to the concept of human rights.

In addition to international organisations and states, NGOs, which mushroomed after the end of the Cold War, have played an ever increasing role in human rights promotion and advocacy. If during the Cold War, as the former UN Secretary General Boutros Boutros-Ghali observed, "it was not possible to have any contact with non-governmental organisations in the Soviet Union, for example, because this would be seen as neo-imperialist intervention",[26] in the post-Cold War era "the decline of oppressive regimes and the rise of democracy has tempered the former automatic hostility by governments toward the activities of local and international NGOs".[27] Today, facilitated by developments in information and communications technology, "NGOs, lobbying for human rights ... have become serious participants in international affairs. Linked with these larger international groups ... are thousands of indigenous NGOs, gathering information on particular issues and pressing their governments to live up to international standards".[28] NGOs now actively participate in setting the agendas for international organisations, state governments and corporations.

With all these developments, as has already been observed by a number of scholars, human rights have become a global norm which plays a causal role in structuring the behaviour of different actors. Thus, Risse, Popp and Sikkink have argued that human rights have emerged as a global standard which increasingly governs the behaviour of individuals and states: "human rights norms ... prescribe rules for appropriate behaviour and help define identities".[29] Evans goes as far as saying that human rights are now hegemonic.[30] Donnelly argues that "regimes that do not at least claim to pursue ... popular political participation

26. Quoted in B. Crossette, "UN Leader to Call for Changes in Peace-keeping", *New York Times* (3 January 1995), p. A3.

27. L. Gordenker and T. Weiss, "Pluralizing Global Governance: Analytical Approaches and Dimensions", in T. Weiss and L. Gordenker (eds.), *NGOs, the UN, and Global Governance* (London: Lynne Rienner, 1996), p. 30.

28. *The Economist*, "The Conscience of Mankind", in a Special Survey entitled "Survey of Human Rights Law" (5 December 1998), p. 3; see also *idem*, "The Power of Publicity", in *ibid.*, p. 13.

29. T. Risse, S. Ropp and K. Sikkink, "The Socialization of International Human Rights Norms into Domestic Practices: Introduction", in T. Risse, S. Ropp and K. Sikkink (eds.), *The Power of Human Rights: International Norms and Domestic Change* (Cambridge: Cambridge University Press, 2001), p. 8.

30. Evans, *op. cit.*, p. 18.

('democracy'), and respect for the rights of their citizens ('human rights') place their national and international legitimacy at risk".[31] In the remainder of this section I will extend these analyses by using Foucault's categories to demonstrate how agents are increasingly evaluated and evaluate themselves with reference to the concept of human rights.

To recall the earlier discussion, Foucault places an especial emphasis on human sciences as vehicles of truth. With respect to human rights, the participation of academics in the global discourse of human rights through publications, teaching activities and political involvement has been very extensive. There is no space here to carry out a review of the academic literature on human rights. It will suffice to note that human rights have been one of the most, if not the most, widely written about subject in the last 20 years or so. Although it would be unfair to say that there is a consensus among academics on the need to promote human rights around the world, for there are many critical works on human rights, we may still acknowledge the fact the dominant majority of scientific publications is concerned with technical-instrumental questions related to the discussion of the means to implement human rights. In such literature the validity of the concept of human rights is not addressed; human rights are assumed as something self-evident and the analysis usually revolves around particular types of rights in specific places, and the means to ensure their respect. In addition to this, different training programmes, university courses and even entire degrees in human rights have mushroomed at universities around the world. It is now hardly possible to undertake a degree in politics, sociology history or International Relations without studying human rights. Furthermore, academics often perform official public missions for international organisations or states, and work for different think tanks that help develop government policies.

With this in mind let us return to the concept of power developed by Foucault. As argued above, we can infer from Foucault's work that his conception of power has two principal features: on the one hand, power is not reducible to agents that exercise it, but mediates the dominant view of what constitutes normal and acceptable agency, and is based on the prevailing conventions which determine how to deal with those who deviate from the norm; on the other hand, Foucault describes forms of power that are not based on repression, but which are positive in that they "produce" behaviour even in the absence of coercion, as agents exercise power over themselves in order to conform to the dominant norm. It is the argument of this article that these two dimensions of power described by Foucault can help us understand better the operation of the global discourse of human rights, the causal power that it has with respect to social practice. This is so because in the existing analyses, particularly in the analyses of the use of force in the name of human rights, the latter is treated in a reductionist manner as a justification invented by Western states (in the case of Realist analyses) or by Western imperialism (in the case of radical critiques) to dominate or colonise the world. Such analyses fail to inquire into the conditions that must be fulfilled for precisely human rights and not some other concept to be invoked as a justification. To put it differently, the question as to what must be the case for policies such as military interventions, the use of sanctions, or the setting of respect for human rights as

31. J. Donnelly, "Human Rights, Democracy and Development", *Human Rights Quarterly*, Vol. 21, No. 3 (1999), p. 608.

a pre-condition for economic aid to be possible, is ignored. Foucault's categories enable us to develop a more holistic analysis which treats human rights discourse as a framework which makes such policies possible and increasingly acceptable. Let us start with the examination of the exercise of power by some agents over others with respect to human rights. In what follows we will examine the following instances of the exercise of power with reference to human rights: Western states with respect to non-Western states; non-Western states with respect to non-Western states; NGOs with respect to business enterprises; and business enterprises with respect to non-Western states.

The 1990s have been described by Mary Kaldor as a "decade of humanitarian intervention".[32] It started with a UN-authorised intervention in northern Iraq codenamed "Operation Provide Comfort" and undertaken by the United States, Britain and France. The objective of the intervention was to rescue Iraqi Kurds who had organised an uprising during the Gulf war and faced violent repression by the government of Hussein. In its Resolution 688 the Security Council condemned "the repression of the Iraqi civilian population in many parts of Iraq" and authorised military action "to ensure that the human and political rights of all Iraqi citizens are respected" (S/RES/0688 (1991)). "Operation Provide Comfort" lasted for three months before it was succeeded by another operation, codenamed "Poised Hammer", which involved maintaining a contingent of troops in Turkey which would be ready to act should Iraq restart the repression of Kurds "threatening to use force against Iraq if human rights are abused".[33]

The Intervention in northern Iraq was followed by military action in Somalia in 1992/93, also authorised by the UN Security Council. Following reports of massive violations of human rights and widespread famine resulting from civil war that had raged in Somalia since January 1991, in December 1992 the UN Security Council authorised the deployment of a US-led force: "Expressing grave alarm at continuing reports of widespread violations of international humanitarian law occurring in Somalia, including ... [the Security Council] authorises the Secretary-General and Member States ... to use all necessary means to establish a secure environment for humanitarian relief operations in Somalia" (S/RES/794 (1992)). Although ultimately unsuccessful, this intervention contributed to the development of the practice of human rights in the post-Cold War period.

Another example of human rights enforcement is "Operation Restore Democracy" in Haiti in 1994, carried out by the United States and authorised by the Security Council. The goal of the intervention was to reinstate the government of democratically elected President Aristide which had been overthrown by the military in 1991. UN Resolution 940 authorised the use of force to stop "the continuing escalation of systematic violations of civil liberties" and to assist in "the restoration of democracy in Haiti" (S/RES/940 (1994)). A US-led force was established, codenamed "Operation Restore Democracy". It consisted of more than 15,000 US troops and a symbolic group of soldiers from other countries, such as Bangladesh, Barbados, Guyana, Ghana and Great Britain. The operation succeeded and President Aristide was returned to power.

32. M. Kaldor, "A Decade of Humanitarian Intervention: The Role of Global Civil Society", in H. Anheier, M. Glasius and M. Kaldor (eds.), *Global Civil Society 2001* (Oxford: Oxford University Press, 2001).

33. *Financial Times*, "Last Allies Pull Out of North Iraq" (16 July 1991), p. 6.

Now, if these interventions were undertaken in accordance with the principles of international law, that is, they were mandated by the United Nations, the NATO action in Kosovo in 1999 was undertaken without UN authorisation. It may be argued that such an action was rendered possible to a large extent by the growth of the global discourse of human rights, of its normative power, and by a growing acceptance at the international level that force may be used to protect human rights. In other words, it was possible for NATO to exercise its power over Yugoslavia with reference to human rights without UN authorisation, and at the same time also without significant international opposition, because in the post-Cold War world the protection of human rights was increasingly seen as legitimate. This constitutes a clear discontinuity with respect to the Cold War period in which the three cases of intervention mentioned above were not justified in terms of human rights but as acts of self-defence. To return to the Kosovo case, in March 1999, in response to Serbian military and police operations, NATO's Operation Allied Force began. The crisis in Kosovo, it was stated, represented "a fundamental challenge to the values of democracy and human rights"; NATO condemned "appalling violations of human rights and the indiscriminate use of force by the Yugoslav government" and asserted that these actions "made necessary and justified the military action by NATO".[34] On 3 June, President Slobodan Milosevic finally accepted peace terms and by 20 June all Serb forces had left Kosovo and security matters in the province had passed into the hands of KFOR, the international peacekeeping force.[35]

Whereas the above cases involved the exercise of power by Western states over non-Western states, the following two cases—the intervention of the Economic Community of West African States (ECOWAS) in Liberia and Sierra Leone—involved non-Western states and their use of human rights as a justification for intervention in other non-Western states.

Thus, in response to massive violations of human rights in Liberia which resulted from the fighting between Charles Taylor's rebel group and the forces of President Doe, ECOWAS sent a military force to Liberia in April 1990, led by Nigeria. This intervention was justified by ECOWAS officials on explicitly human rights grounds: "to stop the abuses of human rights of innocent civilian nationals and foreigners, and to help the Liberian people to restore their democratic institutions".[36] Such a justification was confirmed by the reaction of the Bush administration which praised the willingness of ECOWAS to take the lead in dealing with the Liberian crisis and "stopping massive violations of human

34. NATO, "The Situation in and around Kosovo", Statement issued at the Extraordinary Ministerial Meeting of the NATO Council, Brussels, 12 April 1999, Press Release M-NAC-1(99)51, available: <http://www.nato.int/docu/pr/1999/p99-051e.htm> (accessed 5 July 2009).

35. Owing to constraints of space I do not discuss the War on Terror. However, in the wars in Afghanistan and Iraq human rights rhetoric was present right from the start. Even the attacks of 9/11 were referred to as attacks on human rights. George W. Bush, for example, stated that the "tragedies of September 11 served as a grievous reminder that the enemies of freedom do not respect or value individual human rights. Their brutal attacks were an attack on these very rights", George W. Bush, "President Proclaims Human Rights Day & Bill of Rights Week", The White House Office of the Press Secretary (9 December 2001), available: <http://georgewbush-whitehouse.archives.gov/news/releases/2001/12/20011209.html> (accessed 5 July 2009).

36. Quoted in C. Greenwood, "Is there a Right of Humanitarian Intervention?", *The World Today*, No. 49 (February 1993), p. 37.

rights".[37] In addition to this, virtually all the other African states outside ECOWAS lent their support and "explicitly approved of the intervention on human rights grounds".[38] In November 1992 the UN Security Council retroactively supported the intervention by ECOWAS when it adopted Resolution 788 which imposed an arms embargo on Liberia and acting under Chapter VII endorsed ECOWAS's use of force (S/RES/788 (1992)).

The ECOWAS action in Sierra Leone in 1997 was a response to the overthrow of the democratically elected President Kabbah and again human rights were invoked. ECOWAS stated that its objective was to quash the coup and "to restore democracy and human rights".[39] In March 1998, President Kabbah was returned to office.[40] The ECOWAS intervention in Sierra Leone was also approved retroactively by the UN Security Council. First, in its Resolution 1132 in October 1997, the UN Security Council supported the efforts of ECOWAS and established an arms and oil embargo on Sierra Leone, authorising ECOWAS to enforce it (S/RES/1132 (1997)). Then, on 22 October 1999, it adopted Resolution 1270 which authorised the establishment of a UN Mission in Sierra Leone, with a significant contribution made by ECOWAS "to promote peace and national reconciliation and ... respect for human rights in Sierra Leone" (S/RES/1270(1999)).

Another instance of the way interchangeable agents exercise power with respect to human rights is the campaign of NGOs and different social movements which started in the early 1990s against business enterprises for various malpractices in Third World countries, such as workers' rights abuses, forced labour, employment of child labour, sexual harassment of employees, pollution of host areas and destruction of local communities.[41] The most widely publicised cases were registered in the garment industry, where such giants as Nike, Gap, Adidas and others were reported as severely abusing the rights of their employees, especially at their sweatshops in East Asia.[42] These campaigns succeeded in forcing corporations to abandon the most extreme forms of abuse, and in corporations starting the process of developing codes of good business practice which acknowledged the importance of socially responsible business practice;[43] by the late 1990s virtually all firms had a code of conduct.[44] Another important result of NGO campaigns was putting the operation of multinational corporations (MNCs) on the agendas

37. D. Wippman, "Enforcing the Peace: ECOWAS and the Liberian Civil War", in L. Damrosch (ed.), *Enforcing Restraint* (New York: Council on Foreign Relations, 1993), p. 159.

38. W. Sandholtz, "Humanitarian Intervention: Global Enforcement of Human Rights?", in A. Brysk (ed.), *Globalization and Human Rights* (Berkeley, CA: University of California Press, 2002), p. 214.

39. *The Guardian*, "Nigeria's Intervention Puzzles West Africans" (28 June 1997), p. 17.

40. *The Independent*, "Freetown in Flames as Rebels Retreat" (13 January 1999), p. 11.

41. See I. Manokha, "Corporate Social Responsibility: A New Signifier? An Analysis of Business Ethics and Good Business Practice", *Politics*, Vol. 24, No. 1 (2004), pp. 56–64; *idem*, "Business Ethics and the Spirit of Global Capitalism: Moral Leadership in the Context of Global Hegemony", *Journal of Global Ethics*, Vol. 2, No. 1 (2006), pp. 27–41.

42. S. George, "Corporate Globalisation", in E. Bircham and J. Charlton (eds.), *Anti-capitalism: A Guide to the Movement* (London: Bookmarks, 2001); A. Cockburn, "The Price", *New York Press* (27 September 2001), available: <http://www.nypress.com/14/39/news&columns/wildjustice.cfm> (accessed 5 July 2009).

43. *The Economist*, "Sweatshop Wars" (25 February 1999), p. 4.

44. International Labour Organization, "Overview of Global Developments and Office Activities Concerning Codes of Conduct, Social Labeling and other Private Sector Initiatives Addressing Labour Issues", ILO doc. GB.273/WP/SDL/1.

of different international organisations. For example, the OECD adopted guidelines for multinational enterprises which stated that MNCs must "respect the human rights of those affected by their activities" and "are encouraged to respect human rights, not only in their dealings with employees, but also with respect to others affected by their activities".[45]

Now, what is important to note with respect to business enterprises is that many of them have gone further than just adopting responsible codes of conduct and improving their human rights record and have engaged in policies of human rights defence and promotion. For example, in April 1999, the chief executives of Reebok, Levi Strauss and Phillips sent a joint letter to Jiang Zemin, the President of China, in which they expressed their concern "about the arrest and detention of Chinese citizens for attempting peacefully to organise their fellow workers or to engage in non-violent demonstrations".[46] Another example is given by Anita Roddick of The Body Shop International who was proud to report the successes of the human rights campaign "Make your Mark" organised by The Body Shop and Amnesty International: "we collected over three million thumbprints in 34 countries in support of 12 remarkable human rights campaigners who defend fundamental human rights, often in dangerous and threatening conditions".[47] In 1999 Paul Fireman, CEO of Reebok, wrote a letter to President Habibie of Indonesia urging the release of Dita Sari, a 26-year-old activist who was serving a sentence for non-violent activities in defence of human rights. Fireman explained his action, arguing that Dita Sari's imprisonment "made it difficult for Reebok to honour their commitment to respect the human rights".[48] Another example is the way a number of companies (Adidas, H&M, IBM, IKEA, PepsiCo, Texaco) responded to human rights violations in Burma by withdrawing from the country or ceasing all business transactions with the regime. Finally, corporations have also established partnerships or networks, together with other firms (usually working in the same sector), to promote various rights. For example, BAT (British American Tobacco), along with other tobacco-producing firms, established an organisation called Eliminating Child Labour in Tobacco Growing, and has contributed to the elimination of child labour on tobacco plantations in states where it operates. Hewlett-Packard, as a member an inter-firm organisation known as Business Leaders' Initiative in Human Rights, has financed policies of human rights promotion in different countries of the world.

In other words, these examples provide another illustration of the substitutability of subjects and objects of power (corporations have been both), as well as of the second dimension of Foucault's conception of power, namely its positive dimension, the way dominant discursive structures "produce" behaviour and how agents, even in the absence of coercion, may decide to transform themselves. Once again, this is not to say that corporations now indeed do care about human rights; it is only to highlight the fact that they find it necessary to claim that they do, and to engage in practices that help promote human rights, and thereby contribute to the further development of the global human rights

45. OECD, *OECD Guidelines for Multinational Enterprises* (Paris: OECD, 2008), pp. 14, 39.

46. Quoted in J. Bennett, "Multinational Corporations, Social Responsibility and Conflict", *Journal of International Affairs*, Vol. 55, No. 2 (2002), pp. 401–402.

47. A. Roddick, *Business as Usual* (Wellingborough: Thorsons, 2000), p. 176.

48. P. Fireman, "Business Must Speak up for Human Rights", *Asian Wall Street Journal* (19 August 1999), available: <http://208.55.16.210/Fireman.htm> (accessed 5 July 2009).

discourse. Let us now look at some more examples of this positive aspect of power, namely how non-Western organisations, previously opposed to the idea of human rights, now develop different human rights initiatives and mechanisms. The case of international non-Western organisations provides stronger evidence than the case of individual states, for in the latter case the decision to acknowledge the adherence to human rights principles may be motivated by some immediate considerations (obtaining foreign economic aid, succumbing to pressure from some other actors, etc.), whereas summits of several member states are generally free from such considerations. Moreover, such gatherings provide an opportunity to collectively express a shared opposition or disagreement about different aspects of global politics. The fact that the following non-Western international organisations not only do not express any critique of human rights as they did during the Cold War but increasingly declare their commitment to human rights demonstrates that the second aspect of Foucault's notion of power—the way in which agents exercise power over themselves—is at play.

Thus, the African Union (AU, formerly the Organization of African Unity), beginning in 1999, has placed very strong emphasis on the promotion of human rights in Africa. At a Ministerial Summit in Mauritius in 1999 it was declared that "human rights constitutes one of the fundamental bedrocks on which development efforts should be realised".[49] To recall, during the Cold War, the position of African states was that the absence of human rights (political stability guaranteed by non-democratic regimes) was the necessary condition for development. It was also declared that "Human Rights are universal and indivisible" and that governments must "give parity to economic, social and cultural rights as well as civil and political rights".[50] Thus, civil and political rights were explicitly recognised as human rights. These commitments have been reiterated at numerous subsequent summits and conferences organised by the AU. For example, in 2000 at Lomé, Togo, the new Constitutive Act of the AU was adopted and made human rights one of the organisation's priorities. In 2003, AU officials claimed that "the primary objective of the Union translates into a wide range of goals which include ... the amelioration of democratic process and observance of human rights".[51] The AU also established a special organ called the Peace and Security Council, in order to, among other things, develop "strong democratic institutions ... the observance of human rights and the rule of law".

Another non-Western organisation that has undertaken initiatives with respect to human rights is the Association of Southeast Asian Nations (ASEAN). Thus, in 1993 at a Ministerial meeting in Singapore its Member States declared their "commitment to human rights and fundamental freedoms" and decided to "coordinate a common approach to the application, promotion and protection of human rights".[52] In 1995 the organisation, via a body called the Human Rights Standing

49. African Union, "Grand Bay Declaration and Plan of Action", 12 April 1999, available: <http://www.achpr.org/english/declarations/declaration_grand_bay_en.html> (accessed 5 July 2009).

50. *Ibid.*

51. A. Essy, "Address Delivered by Amara Essy, Chairperson of the Commission of the African Union, Second Ordinary Session of the Assembly of Heads of State and Government", Maputo, 10 July 2003, available: <http://www.africa-union.org/latest%20news/Open%20Speech%20of%20the%20Interim%20Chairperson.pdf> (accessed 5 July 2009).

52. ASEAN "Joint Communiqué: The 26th ASEAN Ministerial Meeting", Singapore 1993, available: <http://202.154.12.3/old/general/ads1993/26amm_93.htm> (accessed 5 July 2009).

Committee, organised a series of meetings among representatives of national human rights institutions, parliamentary human rights committees, and human rights NGOs in the region to discuss proposals for a new human rights mechanism in Southeast Asia. At all subsequent summits ASEAN has reiterated its determination to promote human rights in the region.

Similar declarations and commitments have been made by other non-Western organisations (e.g. ECOWAS, which, as we saw above, played an important role in the development of the practice of human rights enforcement; the SADC (South African Development Community); and Mercosur in Latin America). To reiterate, these initiatives have hardly produced any significant results in terms of human rights promotion and protection. However, they illustrate important changes in the position of non-Western states regarding human rights. From being opponents of human rights (or, to be more precise, of their Western conception as individual political and civil rights), they have transformed themselves into being their advocates. This constitutes a definite historic break or discontinuity, and can be explained by the development of the global discourse of human rights which has created a structure within which human rights has become a standard with reference to which agents are evaluated and evaluate themselves, increasingly finding it necessary to adhere, at least formally or officially, to this standard. And by doing so, they contribute to the further development of this discourse and of its power.

To sum up, we have identified a historical discontinuity, and have seen how in a new discursive structure power is exercised. A strictly Foucauldian analysis would have ended here. However, as already mentioned above, the limitation of Foucault's method is that it does not seek to explain changes. It also separates discourses from the economy, from the functioning of production relations. In the next section we will see how a political economy approach may complement a Foucauldian analytical framework and thereby render the analysis more complete.

A Political Economy Approach to the Global Discourse of Human Rights

As Lecourt put it in his critique of Foucault, Foucault is a theorist of the superstructure "condemned to silence over the link between ideology and the relations of production".[53] In this final section we will use the concept of ideology and will discuss the global discourse of human rights with respect to the relations of production. The argument focuses on the historical relationship of mutual causality between social structures of production on the one hand, and the notion of individuals' rights on the other. The nature of this relationship is analysed with the help of the concept of "ideology", used as a critical concept, or ideology in the negative sense.[54]

The term "ideology" has generally been used to refer to political programmes of social classes or political parties, and this is a meaning which Jorg Larrain

53. D. Lecourt, *Marxism and Epistemology: Bachelard, Canguilhem and Foucault* (London: New Left Books, 1975), p. 207; see also N. Poulantzas, *State, Power, Socialism* (London: New Left Books, 1978).

54. J. Larrain, *Marxism and Ideology* (London: Macmillan, 1983); J. Maclean, "Belief Systems and Ideology in International Relations: A Critical Approach", in R. Little and S. Smith (eds.), *Belief Systems and International Relations* (Oxford: Blackwell, 1988); P. Allman, *Critical Education against Global Capitalism: Karl Marx and Revolutionary Critical Education* (London: Bergin & Garvey, 2001).

designates as "positive".[55] However, there is another usage of the term, namely what Larrain calls "ideology in the negative sense", which Marx, particularly in *The German Ideology*, uses as a critical concept to refer to "ideas which express practice inadequately".[56] For Marx, "ideology is conceived of as a distorted form of knowledge"[57] or as "a defective way of thinking".[58] Such a "distorted form of knowledge" is not reducible to a deliberate project of a class; ideological ideas are "*encouraged* to a great extent"[59] by a social reality which is "inverted", that is, which appears different from what it really is and favours the development of ideas that express it inadequately, and which may be shared by all social classes. However, ideology necessarily serves the interests of those social forces that benefit from the existing order, for the role it plays in the negation or concealment of contradictions inevitably contributes to the reproduction of those contradictions.[60] We will see below what form ideology so defined assumes in pre-capitalist societies, how the emergence of the notion of individual rights with the rise of capitalism may be said to have led to the development of a new ideology specific to capitalism, and what causal powers and liabilities this ideology has in the context of global capitalism.

Thus, the feudal relations of production were characterised by various forms of personal/juridical dependence of producers on those appropriating the surplus of their labour, and the function of surplus extraction was carried out by means of political coercion. It was a mode of production dominated by the land and a natural economy, in which neither labour nor the products of labour were commodities. The peasants who occupied and tilled the land were not its owners, with agrarian property being privately controlled by feudal lords, who extracted surplus from the peasants by politico-legal relations of compulsion in the form of labour-service, in kind or in the form of money.[61] Feudal society was thereby characterised by a juridical amalgamation of economic exploitation with political authority—property was "politically constituted".[62] The feudal lord, in his turn, would often be the vassal of a feudal superior, and the chain of such dependent tenures (linked to military service) would extend upwards to the peak of the system—a monarch. Political sovereignty was not focused on a single centre but was divided into particularist zones with overlapping boundaries, and no universal centre of competence.[63]

The direct political exploitation of producers and unequal class relations of the feudal order had an ideology which helped to sustain them. This ideology involved interpreting social relations as having a divine origin and purpose with each individual performing a specific function ascribed by God: every member of society must receive the means suited to his or her function, and must claim no more. This ideology spiritualised the material by incorporating it in a divine universe,[64] or, as Larrain puts it, in feudalism

55. Larrain, *op. cit.*, p. 4.

56. *Ibid.*, p. 23.

57. Maclean, *op. cit.*, p. 309.

58. Allman, *op. cit.*, p. 47.

59. G. Lukács, *History and Class Consciousness* (London: Merlin Press, 1971).

60. Larrain, *op. cit.*, p. 28.

61. B. Hindness and P. Hirst, *Pre-capitalist Modes of Production* (London: Routledge, 1975), p. 223.

62. R. Brenner, *Merchants and Revolution: Commercial Change, Political Conflict and London's Overseas Traders, 1550–1653* (Princeton: Princeton University Press, 1993).

63. P. Anderson, *Passages from Antiquity to Feudalism* (London: Verso, 1996), p. 148.

64. R. Tawney, *Religion and the Rise of Capitalism* (London: Penguin, 1990), pp. 35–36.

> ideology assumes a religious form; the justification of personal dependence is found in a sacred order which is revealed by God and which consequently cannot be altered by man. Personal dependence upon, and loyalty to, the landlord is spontaneously expressed in the ideological submission to God, from which all subordination is modeled.[65]

As one Medieval English hymn verse had it, "The rich man in his castle / The poor man at his gate/God made them, high or lowly/And ordered their estate".[66] Although such ideological interpretation of social relations was shared by all classes, they necessarily benefited the ruling classes whose dominant position was seen as an outcome of a divine will. The notion of individual rights would have been inconceivable in the feudal social setting in that it would have contradicted its economic foundation—the relations of production and the manner of the extraction of surplus.

However, beginning with the late 16th and early 17th centuries, profound transformations of social relations took place in Europe, starting in England and later spreading to the rest of Europe. What is characteristic of the emerging order is the central role of the market: capitalism is a system in which goods and services are produced for profitable exchange and all economic actors are dependent on the market—workers, who must sell their labour power for a wage, and capitalists, who depend on the market to buy their inputs and to sell their outputs.[67] The function of surplus extraction is gradually transformed from being based on politico-juridical means into one carried out in the market by economic means—through a wage–labour relation—as the producer is separated from the conditions of labour and the appropriator has absolute private property in the means of production. Once producers lose free access to the means of production, labour becomes a commodity, "that very special commodity that makes capitalism possible".[68] Thereby feudal politico-juridical inequality and coercion become unnecessary for surplus extraction since it is now the economic need that compels the worker to transfer surplus labour to the capitalist in order to gain access to the means of production. In other words, the social allocation of resources and labour no longer takes place by means of political direction or religious obligation, but through the mechanisms of commodity exchange.[69] This gradually makes possible political emancipation and political equality, as political coercion is no longer necessary for surplus to be extracted from the direct producers.[70] It allows for the development of new political structures breaking with the tradition of divine right and emphasising the equality of individuals, their natural rights and their power to establish and change political authority. However, what is crucial is that political and civil liberty and equality are increasingly seen as *the* liberty and *the* equality, while the fact that exploitation and inequality exist in the market is taken for granted and objectified. In other words, a new form of ideology

65. Larrain, *op. cit.*, p. 38.

66. M. Barrett, "Ideology, Politics, Hegemony: From Gramsci to Laclau and Mouffe", in S. Žižek (ed.), *Mapping Ideology* (London: Verso, 1994), p. 253.

67. E.M. Wood, *The Origins of Capitalism: A Longer View* (London: Verso, 2002), p. 2.

68. K. Marx, *Capital*, Vol. 1 (London: Penguin, 1976), p. 270.

69. E.M. Wood, *Democracy against Capitalism: Renewing Historical Materialism* (Cambridge: Cambridge University Press, 1995), pp. 28–29.

70. Marx, *op. cit.*, pp. 270–271.

develops constituted in the view that in capitalism individuals are free and equal, freedom and equality being in fact confined to the realm of the political. Such an ideology first appears in the writings of social contract theorists such as Hobbes and Locke, while the foundations of its internationalisation are laid by Kant.[71] As Rupert puts it,

> capitalism's structural separation of the economic from the political may have crucial ideological effects: it enables the wage relation to take on the appearance of a voluntary exchange between abstract individuals in the market; while, at the same time, the state may appear as a class-neutral public sphere in which abstract individuals may interact as formally equal citizens pursuing an instrumental politics of self-interest.[72]

Thus, capitalism "makes possible a form of democracy in which formal equality of political rights has a minimal effect on inequalities or relations of domination and exploitation in other spheres".[73] The notion of individual rights is not only compatible with the functioning of the market but also codifies some of its most fundamental elements—the right to property and the right to employment objectify the existence of private property and wage labour which constitute the key institutions of capitalism.

The ideology of freedom and equality embodied in the notion of rights has developed historically in an organic relationship with capitalism. There is no space here to examine this historical development in detail. What I wish to emphasise is that changes in the relations of production—from feudal to capitalist—must be examined to understand the emergence of the notion of natural rights. This is not to say that economic relations determine other social institutions; however, they create a setting within which certain non-economic forms are rendered possible, while others are impossible or inconceivable.

With the end of the Cold War and the collapse of the socialist alternative, the development of global capitalism rendered possible the development of the global human rights discourse. Again, the latter was not necessitated by the former; however, a global setting more conducive to the globalisation of human rights appeared with the transition of the former socialist states to capitalism. The direct causal role was played by a series of different conjunctural factors: the fact that the victory of the West in the Cold War was interpreted as the victory of freedom, which was now to be spread elsewhere; the breakthroughs in information and communications technology which made human rights abuses more difficult to conceal; and the growth of a global civil society now free from Cold War constraints.

Now the historical relation between individual rights and capitalism raises one important question: what is the nature of this relationship today in the era of global capitalism? I suggest that the global discourse of human rights today plays the role of Gramsci's moral leadership in the context of the development of a hegemonic order at the heart of which lie the interests of transnational

71. I. Manokha, *The Political Economy of Human Rights Enforcement* (London: Palgrave, 2008), pp. 86–93.

72. M. Rupert, *Ideologies of Globalization: Contending Visions of a New World Order* (London: Routledge, 2000), p. 3.

73. Wood, *Democracy against Capitalism*, *op. cit.*, p. 224.

capital. In the remainder of this section I will develop this argument in more detail.

Gramsci distinguishes between a rule based on coercion and a rule based on consent and "intellectual and moral leadership"[74] or hegemony. He argues that consent is secured by dominant social forces by means of education of subordinate groups to ensure that they share a particular world outlook and a particular set of beliefs on the one hand, and the universalisation of certain moral values and norms on the other. For Gramsci, the task of hegemonic forces is to "educate" the masses, to "produce" subordinate groups' understanding of their role in society and of society itself, to make sure that they accept the existing society "without criticism".[75] Such a "production" of consent must be complemented by a system of moral values which, once established, influences the subordinate group's "moral behaviour and the direction of [their] will".[76] In short, the establishment of a hegemony consists in "building a new intellectual and moral order, and hence the need to elaborate the most universal concepts".[77]

Gramsci's concept of hegemony has been applied to the realm of the international by a number of scholars who are referred to as the "neo-Gramscians". They share the view that the last three decades have constituted a phase of transition from a post-war US hegemony to a new form of hegemonic formation centred on transnational capital.[78] The hegemonic position of highly mobile capital is institutionalised in a variety of multilateral organisations and treaties under pressure not only from transnational corporations but also "organic" intellectuals such as corporate allies in governments, inter-state organisations, and different policy-making and advisory networks, which develop neoliberal policy guidelines and reduce the role of states to ensure the working of the market mechanism. Gill argues that at the core of the emerging hegemonic formation is a nucleus which comprises the following elements: G8 state apparatuses, transnational capital, associated privileged workers and smaller firms (businesses linked as contractors or suppliers, import–export firms, and service companies such as stockbrokers, accountants, consultancies and lobbyists).[79] Competitiveness in the world market has become the ultimate criterion of state policy which justifies the gradual removal of the measures of social protection built up in the era of Keynesianism and the welfare state,[80] and a form of "new constitutionalism" or the institutionalization of neo-liberalism at the macro-level of power in the quasi-legal restructuring of the state and international political forms[81] has developed. The element of consent in the developing global order, despite international civil society protests, is underlined by a relative decline

74. A. Gramsci, *Antonio Gramsci: Selections from Prison Notebooks* (eds. Q. Hoare and G. Nowell Smith) (London: Lawrence & Wishart, 1971), p. 57.

75. *Idem, The Modern Prince and Other Writings* (New York: Lawrence & Wishart, 1957), p. 66.

76. *Ibid.*, p. 67.

77. *Ibid.*, p. 5.

78. For more on this see Manokha, *The Political Economy of Human Rights Enforcement, op. cit.; idem*, "Al-Qaeda Terrorism and Global Poverty: New Social Banditry", *Journal of Global Ethics*, Vol. 4, No. 2 (2008), pp. 95–105.

79. S. Gill, "Globalization, Market Civilization, and Disciplinary Neoliberalism", *Millennium*, Vol. 24, No. 3 (1995), pp. 400–401.

80. R. Cox, "Civil Society at the Turn of the Millennium: Prospects for an Alternative World Order", *Review of International Studies*, Vol. 25, No. 1 (1999), p. 12.

81. Gill, *op. cit.*, p. 412.

of class and radical politics, and a noticeable decline in the power and activity of trade unions as enterprises increasingly locate their plants in different countries which makes trade union organisation and bargaining more difficult.[82]

The role of moral leadership in the development of consent has so far been neglected in neo-Gramscian analyses, and yet it is emphasised repeatedly by Gramsci. In addition to the arguments quoted above with respect to the need to create "a new moral order", the role of moral leadership consists in contributing to the reproduction and objectification of the existing relations of production, for "although hegemony is ethical-political, it must also be economic, must necessarily be based on the decisive function exercised by the leading group in the decisive nucleus of economic activity".[83] From these observations by Gramsci we may infer that moral leadership has two principal characteristics: on the one hand, it involves universalisable moral values, and on the other it naturalises, via these values, the existing order and its material foundations so that they are taken for granted or attain the status of "common sense".[84]

The global discourse of human rights possesses both of these dimensions. On the one hand, it has a very strong moral import. The struggle for freedom, for the liberation of the oppressed, their rescue from the hands of torturers and the persecution of the perpetrators are by their very nature morally charged actions and are usually accompanied by a significant degree of moral rhetoric. On the other hand, these moral values contribute to the objectification of the key mechanisms of the global economy, their being taking for granted and naturalisation. Now this claim needs to be substantiated, for the discourse of human rights also includes social and economic rights, which are seen by a number of analysts as having counter-hegemonic potential. It may be argued that, indeed, these rights presuppose certain constraints on the operation of capital, and a certain degree of protection of workers. However, this potential is yet to be realised for it is the political and civil rights that prevail by far in the global discourse of human rights, as has been observed by numerous commentators. Furthermore, the counter-hegemonic potential of social and economic rights needs to be treated with caution, for some of these rights not only fail to question the existence of global capitalism but also naturalise one of its most fundamental mechanisms—the wage–labour relation—by declaring such rights as the right to work, to fair wages, to equal opportunity for everyone to be promoted in his/her employment and to periodic holidays, with pay, as universal human rights. As things stand today, the present article argues that the global discourse of human rights, dominated by negative rights, contributes to the development of Gramsci's quality of moral leadership in the context of the developing hegemony of capital.

Conclusion

This article has attempted to analyse the global discourse of human rights from a Foucauldian perspective. It has been demonstrated that this discourse has

82. See Cox, *op cit.*; see also U. Beck, *The Reinvention of Politics: Rethinking Modernity in the Global Social Order* (Cambridge: Polity Press, 1997).

83. Gramsci, *Antonio Gramsci, op. cit.*, p. 161.

84. *Idem, The Modern Prince and Other Writings, op. cit.*, p. 67.

contributed to making human rights a kind of a global standard or norm with reference to which agents are increasingly evaluated and evaluate themselves. Using two dimensions of Foucault's conception of power—the substitutability of subjects and objects of power who do not possess but exercise power with reference to the prevailing standard of behaviour, as well as the manner in which power may be exercised by agents over themselves even in the absence of external coercion—we have discussed the way in which global discourse of human rights has conditioned the behaviour of different forms of agency. We have seen how different actors have exercised power over those who were seen as not conforming to the norm: Western states with respect to non-Western states; non-Western states with respect to non-Western states; NGOs with respect to business enterprises; and business enterprises with respect to non-Western states. We have also discussed the way in which some entities—most notably business enterprises and non-Western states—have increasingly found it necessary to declare their adherence to human rights norms and to develop mechanisms and policies for their promotion, without any coercion from other actors. Even if in many cases such declarations and policies fail to produce any substantial change in terms of respect for human rights and amount to only formal commitments, it is nevertheless important to explain why these agents have suddenly sought to make such declarations and develop human rights policies. It has been argued that Foucault's analytical framework helps us understand this by designating positive forms of power, that is, power as different from repression or negation, as a productive or creative force, and by helping us to explain how dominant discursive structures may lead to changes in the behaviour of some agents. It has also been argued that by making human rights commitments and by engaging in human rights practices, agents contribute to the further growth and power of the global discourse of human rights.

A purely Foucauldian analysis would have left it there. However, as has been argued in the final section, such an analysis of the global human rights discourse would have been incomplete. It would have treated human rights in isolation from the economy, from the relations of production, and it would have left their historical development and their rapid ascendance in the post-Cold War period unexplained. The last section of the article argues that the notion of individual rights emerged with the development of capitalism and has developed in an organic relationship with it. With the help of the concept of ideology in the negative sense its causal role with respect to the development of capitalism has been described as residing in the ideological representation of political and civil liberty and equality as the liberty and inequality naturalising and objectifying the relations of power and inequality in the economy. It has then been argued that at the current historical juncture, the causal role of global human rights discourse resides in providing transnational capital with a Gramscian quality of moral leadership in the context of its developing hegemony.

It seems that a purely Foucauldian analysis of the global discourse of human rights would have ended on a positive note. After all, this global discourse contributes to the promotion of important values and ideals which are increasingly difficult to disregard without facing some form of opposition or reaction. The political economy approach that has been used here to complement the Foucauldian framework allows us to take a more critical stance towards this discourse because of its internal relationship with the capitalist production structure. The global discourse of human rights may be said to produce an unintended consequence of contributing to the growing power of capital over the state/society complex. As such, this noble idea turns out to play a role in the reproduction of different forms of human misery and suffering affecting millions of people around the world that result from the current operation of the global economy.

Hobbes, War, Movement

LEONIE ANSEMS DE VRIES and JORG SPIEKER

While informed by Foucault's understanding of power in terms of war and circulation, this article challenges Foucault's static reading of Hobbes. Contextualising Hobbes's political thought within the scientific ideas that he was inspired by, this article reveals that there is more to Hobbes than the static, depoliticising image of the contract. Hobbes's political thought is premised upon an ontology of movement; that is, his account of political order pivots on a double movement in which war constitutes the very possibility of social and political relations as well as of their continued reproduction via circulation. It is this conceptualisation of order that makes Hobbes's liberal political thought genealogically significant. And it is the model of a play of (re)productive movements—in certain respects close to Foucault's own conception of power—that can be used productively for thinking about governance and resistance today.

Introduction

Hobbes's idea of the movement from the state of nature to political society, from the state of war to a condition of peace, is one of the most influential motifs shaping our modern political imagination. Clearly, this image conceals more than it reveals with regard to the origins of political order, the relationship between peace and war, and the functioning of power in modern societies. To conduct a critique of liberalism is thus always also to challenge this motif and the problematic distinctions that it helps to maintain, including that between reason and irrationality or that between order and disorder. In this paper, we seek to challenge the conceptualisation of politics in terms of a movement from the state of nature to political society through Hobbes himself. We will suggest that Hobbes's political theory also contains a different model of movement—one that locates him at the foundation of liberal thought, and one, moreover, that we can still employ today for thinking critically about (liberal) politics.

The point of departure for our engagement with Hobbes is Michel Foucault's conceptualisation of power. In this paper we will consider what are arguably two of the most significant notions that appear in Foucault's critical political thought: war and circulation. In *Society Must be Defended*, Foucault presents a concept of war that exceeds the battlefield sense in which the concept is usually understood and therefore ambiguates the distinction between war and peace. The idea that political order is always a continuation of war by other means runs directly counter to the key assumption of liberal political thought, namely that there is something like a transition from a state of war to a condition of peace. In *Security, Territory, Population*, Foucault moved from the problem of war to the emergence and development of a distinctly liberal governmentality. One of the notions that emerges from this analysis is that of circulation. For Foucault, circulation is the space of operation of, and the principle of organisation for, modern biopolitics and hence of critical importance for understanding the workings of liberal regimes of power. Both the notion of war and that of circulation have become central to Foucauldian analyses of liberal regimes.

For Foucault, the figure of Hobbes represents a juridical view of power based on the theory of sovereignty and the idea of the contract. If Foucault himself understands power in terms of war, biopolitics and circulation then Foucault's Hobbes represents a view of power in terms of peace, sovereignty and the law. Central to Foucault's Hobbes is the theory of the contract and the motif of the transition from the state of nature to civil society. While largely in agreement with this reading of Hobbes we forward the following hypothesis: if Hobbes is the man who formulated this motif most influentially he is also the man who formulated the political problem to which it represents a solution; and if his solution has come to serve liberal political discourse by reifying the workings of power, then his articulation of the original problem can still help us to think critically about liberal regimes and the ways in which power functions therein. Our intention, in other words, is not so much to reject and to advance an alternative as such to, but to accept and to develop in alternative directions, Foucault's reading of Hobbes.

Whilst Foucault's encounter with Hobbes is marked by an essentially static and peaceful transition from state of nature to the state of order, we approach Hobbes via the notion of movement. Contra Foucault, we will explore the centrality of war and circulation in Hobbes's thought. Moreover, in a second departure from Foucault, we will suggest that these movements are both inextricably linked and should be understood at the level of ontology. There are two movements in Hobbes. First, there is a field of disruptive, disorderly and clashing forces, or the radical uncertainty and indeterminacy of the human condition, in which claims to truth, knowledge, power, etc. remain undecided, yet which produces the very possibility of the determination of such claims. The second is a movement of circulation, which seeks to reproduce and manage the order distilled, marked out and bounded by rendering orderly, legitimate and rational the disruptive field. These two movements, productive and reproductive movement, simultaneously and continuously reinforce and undermine one another, thus constituting a double movement. Hobbes's political theory, in other words, is characterised by a play of (re)productive movement(s).

The first section of this paper problematises Foucault's Hobbes by engaging with Foucault's writings on war and circulation. From this discussion, Hobbes emerges as the theorist who sought to obfuscate war and instil movement. We will introduce the move beyond Foucault's Hobbes by briefly considering two alternative interpretations of Hobbes's political thought. Emerging from this engagement with George Kateb's and William Connolly's readings of Hobbes is a line of thought that emphasises the centrality of movement in Hobbes. In order to explore and critique this perspective via Hobbes himself, the second section provides a genealogical sketch of the developments in scientific thinking which profoundly influenced Hobbes's understanding of both life and politics. In his writings, we can find the influence primarily of two theorists of movement:

Galileo and Harvey. Here, Hobbes emerges as the theorist of movement and circulation; the theorist for whom circulation is not merely instrumental, but ontological. Circulation constitutes the principle of order(ing) in respect to the human body as well as the body politic. Via an examination of the nature of the state of nature in Hobbes, and the movement to a state of order, the third section draws out the double movement that, we suggest, marks Hobbes's political theory: (re)productive movement. From this discussion, Hobbes emerges as the theorist who indeed—as Foucault excellently demonstrates—sought to obfuscate war, yet, and here we depart from Foucault, for whom war is at once key to his thought. War constitutes the very possibility of social and political relations, and of their continued reproduction via circulation; and war threatens to disrupt this order(ing) of circulation ad infinitum. This complex, perpetual play of ordering and disordering (re)produces, defines, marks out and undermines political order.

In conclusion, we will contend that the notion of a play of (re)productive movement(s) can be used productively for thinking about (global) modes of governance, politics and resistance today. Through the identification of movement at the basis of his political theory, Hobbes emerges as a thinker whose genealogical significance to liberal political thought is profound.

Foucault's Hobbes

Some of Foucault's most insightful and significant contributions to our understanding of liberal regimes of power have emerged in the lecture series *Society Must be Defended* and *Security, Territory, Population*. In each course, Foucault develops his hypotheses through a key distinction between his own conceptualisation of power on the one hand, and a more conventional approach to power on the other. Whereas *Society Must be Defended* is largely conducted through an opposition between peace and war, *Security, Territory, Population* can be said to rely partly on the opposition between biopolitics, population and circulation on the one hand, and sovereignty, territory and the law on the other. For Foucault, the figure of Hobbes is predominantly positioned on the side of those more conventional conceptualisations of power that Foucault sought to transcend. That is, for Foucault, Hobbes is a theorist of peace and law rather than of war and circulation. What we will suggest is that Foucault's reading of Hobbes relies on the theory of the contract and that this is too restricted a reading of Hobbes. On this basis, we will point towards an alternative reading of Hobbes—one that moves beyond the logic of the contract and is therefore able to appreciate the notions of war and circulation in his thought.

Society Must be Defended is an insightful and provocative study of the emergence and development of a conceptualisation of power that takes war as its grid of intelligibility. Rather than simply embracing this conceptualisation, which is his own, Foucault is interested in ambiguating it. And while sometimes overlooked by commentators, this element of self-critique is of profound importance to critical political thought. What Foucault reveals, at the level of discourse, is the fine line that separates forms of resistance from the regimes of power they seek to counter and often come to resemble in the course of this endeavour. But in the process of ambiguating certain distinctions, Foucault unavoidably makes distinctions of his own. The theory of sovereignty embodied by the figure of Hobbes appears as the primary conceptual opposite against which Foucault defines his own efforts. Foucault's endeavour to reveal what is concealed by the juridical model of sovereignty is guided by the model of war through which we can make visible and reflect upon the continuous reproduction of those relations of power and domination that ultimately sustain political and social order. According to this conception, which Foucault sets out to trace genealogically, political power constitutes a continuation of war by other means. Rather than putting an end to war and establishing peace through a contractual agreement, political power conducts a permanent "silent war" by re-inscribing a particular relationship of force in the institutions of society and the bodies of individuals.[1] It is important to note that Foucault first articulated this model in the first volume of *The History of Sexuality* in terms of a "field of force relations":

> . . . it is a question of orienting ourselves to a conception of power which replaces the privilege of the law with the viewpoint of the objective, the privilege of prohibition with the viewpoint of tactical efficacy, the privilege of sovereignty with the analysis of a multiple and mobile field of force relations, wherein far-reaching, but never completely stable, effects of domination are produced. The strategical model, rather than the model based on law.[2]

Thus emerges the key genealogical distinction between a philosophico-juridical discourse of peace on the one hand, and a historico-political discourse of war on the other. Foucault's twin aim will be to trace the genealogical development of the discourse of politics-as-war from its beginnings in 16th- and 17th-century revolutionary counter-state formations, and to show how this discourse came to be appropriated, eventually, by the state itself so as to formulate a modern biopolitical security discourse.

Foucault needs to establish the distinction he suggested between a discourse of war and a discourse of peace. And it is in this context that Foucault also presents what is undoubtedly his most intensive engagement with the political theory of Hobbes, whom he considers to be the main representative of the discourse of peace. As soon as we start thinking about power and war, Foucault asserts, the name of Hobbes springs to mind.[3] That Hobbes is, in fact, not at all the thinker who conceptualises politics as war is what Foucault seeks to show. Hobbes, Foucault suggests, is not the man for whom everything begins with war, for the Hobbesian state of war is not a historically specific war in the sense of battle, blood and victory.[4] Instead, the "war of all against all" is based on equality and hence ending in victory or in a relationship of domination is impossible. Foucault argues that the differences that characterise Hobbes's state of war are minor differences and that these minute variations of individual strength do not lead to actual war but to a state of "unending diplomacy" or "relationships of fear".[5]

Foucault then asks how this state of war gives birth to sovereignty and rightly emphasises that even in what Hobbes calls "sovereignty by acquisition" the

1. Michel Foucault, *Society Must be Defended* (London: Penguin, 2003), p. 16.
2. Michel Foucault, *The History of Sexuality I* (London: Penguin, 1998), p. 102.
3. Foucault, *Society Must be Defended*, *op. cit.*, p. 59.
4. *Ibid.*, p. 92.
5. *Ibid.*

determining factor leading to the establishment of sovereignty is not defeat in war but fear.[6] Rather than being "the motor behind political order", Foucault concludes, war in Hobbes is "a clash of presentations"; the state of war is in fact "nonwar":

> Hobbes turns war, the fact of war and the relationship of force that is actually manifested in the battle, into something that has nothing to do with war. Basically, Hobbes's discourse is a certain "no" to war. It is not really war that gives birth to States, and it is not really war that is transcribed in relations of sovereignty or that reproduces within the civil power—and its inequalities—the earlier dissymetries in the relationship of force that were revealed by the very fact of the battle itself.[7]

The point is that war makes intelligible the sovereign's history, and to escape this coercive history Hobbes calls upon the state of nature—the place where sovereign power hides its "grid of intelligibility". The story of the contract presents political power as unproblematic in its origins and as static in its workings. Foucault is entirely right in his conclusion that Hobbes sought to ban war from the discourse of politics, and it is not our intention to criticise this conclusion per se. What we would like to do instead is to ask the question of whether we cannot make a more insightful point about war in Hobbes by considering his political theory as a whole of which the metaphor of the contract, which Foucault makes central to his reading, is merely one part.

In *Security, Territory, Population*, Foucault shifts emphasis from the problem of war to the emergence and development of a distinctly liberal governmentality. Foucault sets out to trace the genealogical shift from a problematisation of political power centred on the exercise of sovereignty over a territory to one concerned with the regulation of populations. One of the central notions emerging from this analysis is that of circulation. What we intend to show is that Foucault's discussion of the problem of circulation relies on an implicit opposition between liberalism, government and the concern with circulation on the one hand, and the relatively static framework of sovereignty and the figure of Hobbes on the other.

What is important to note with regard to Foucault's understanding of circulation is that it is part of a general shift of emphasis rather than of a simple transition from territory to population. The important question is therefore not so much one of *whether* the idea of circulation is either present or absent within a particular political rationality but of *how* the notion of circulation is present therein. What we need to look at, according to Foucault, is "the general economy of power" in order to identify whether, in what way, and to what extent it displays a focus on territory, on discipline or on the regulation of a milieu and circulation.[8] That this is the appropriate way of looking at circulation becomes apparent in Foucault's reflection on the issue of town planning. What Foucault seeks to understand through the example of town planning is the treatment of space particular to different forms of power. Thus, he refers to three different examples which serve to show the different treatments of space by sovereignty, discipline, and security.

6. *Ibid.*, p. 95.
7. *Ibid.*, p. 97.
8. Michel Foucault, *Security, Territory, Population* (Basingstoke: Palgrave Macmillan, 2004), p. 30.

The reflection on the town in terms of sovereignty is characterised by the primacy of the relationship between sovereignty and territory. The state is constructed like an edifice and the main problem is that of situating the capital city within this edifice. But, crucially, the sovereign is also concerned with the circulation of ideas, orders and commerce—that is, with political and economic circulation around a sovereign centre.[9] In fact, Foucault speaks of a combination of the "old idea" of sovereignty with the "modern idea" of circulation.[10]

This combination appears again in lecture 10 of the course where Foucault looks at reason of state in relation to the problem of obedience. Foucault's text of choice is Francis Bacon's *Of Seditions and Troubles*. For Foucault, what is remarkable about this text is first of all the fact that sedition figures as an entirely normal or natural phenomenon.[11] For Bacon, the causes of and the remedies for the problem of sedition are to be found in the belly and in the head—that is, in the domain of the economy and in the domain of public opinion.[12] What emerges from this discussion is that, for Bacon, and for reason of state more generally, sedition is something that needs to be governed; and it needs to be governed through circulation—the circulation of opinion and perception as well as economic circulation. Within the logic of reason of state, the problem of sedition was inseparable from the more generic concern with the circulation of commerce, money and wealth. Reason of state was also interested in modifying people's opinion and their ways of acting as economic and political subjects.[13] Government is to act on the action of others or to conduct their conduct; government is always the government of movement and of circulation. What this suggests is that Foucault does, in fact, grant the notion of circulation a central place in 17th-century political thought. The Classical Age was the era of reason of state and the latter's central element was the police apparatus.

What emerges as one of the main characteristics of apparatuses of security and hence of a distinctly liberal art of government is its conceptualisation of political space in terms of the milieu—that is, in terms of natural and artificial givens or the space of circulation. Foucault points towards the idea that circulation has always been a central problem for government, and that one of the significant differences is that between the naturalness and artificiality of the circulatory space in question. In fact, at the end of the first lecture, Foucault asserts that what we see with the technical problem posed by the town is the "... emergence of the problem of the 'naturalness' of the human species within an artificial milieu".[14] It is in the 18th century, then, that we see the emergence of nature, and the emergence of a natural rather than an artificial space of circulation. Within the political rationality of 18th-century liberalism, the population is considered to have "its own laws of transformation and movement".[15] Importantly, the population, so understood, has internal mechanisms of regulation; it is self-organising and liberal government seeks to manage these self-regulatory processes.

9. *Ibid.*, p. 15.
10. *Ibid.*
11. *Ibid.*, p. 267.
12. *Ibid.*
13. *Ibid.*, p. 275.
14. *Ibid.*, p. 22.
15. *Ibid.*, p. 351.

Foucault makes a clear distinction between theories and doctrines of government such as reason of state on the one hand, and the theory of the contract, the problem of sovereignty and Hobbes on the other. In fact, Foucault argues at some stage that "the pre-eminence of the problem of sovereignty, both as a theoretical question and as a principle of organization, was a fundamental factor in blocking the art of government".[16] For Foucault, the relationship between sovereign power and that which it seeks to govern is, in Hobbes, fully explicable on the basis of the logic of the contract. The only form of movement that is important is the exchange of rights—that is, the actual contractual agreement. Foucault's Hobbes is defined essentially through the theory of the contract or what Foucault regards as political philosophy as opposed to governmentality. Hobbes thus appears as a theorist of peace not of war and as a theorist of the law and not of circulation.

In order to move beyond Foucault's interpretation of Hobbes in terms of the theory of the contract, it is useful to consider briefly some alternative interpretations provided by George Kateb and William Connolly. Both Kateb and Connolly treat Hobbes as a more or less liberal thinker without focusing on the theory of the contract.[17] Whilst Kateb turns attention to the emancipatory implications of Hobbes's radical individualism, Connolly emphasises the production of a specific kind of (self-interested) individual, who is partly free, yet must be continuously regulated. In his interpretation of Hobbes, Kateb focuses on the problem of civil war and on the different arguments that Hobbes deploys in order to show that civil war is irrational and unjustifiable.[18] Kateb sets out to analyse these arguments in detail and for the purpose of developing the following claim: Hobbes, insofar as he develops a radical, special individualism, is an emancipator.[19] Hobbes's radical individualism and his emancipatory politics are expressed in the notion of an "unrenouncable right to life".[20] This right strengthens the individual against the claims of Leviathan and can be understood as a radical critique of the militarist, nationalist and absolutist tendencies in Hobbes's thought.

Of particular interest in the present context is Kateb's contention that although one of Hobbes's chief concerns is to ward off rebellion, disorder and sedition, he equally holds that "[t]he talents and energies of the potentially seditious are indispensable, but they must be properly channelled".[21] The very (irrational) force of disruption is not simply negative or destructive but can, if properly regulated, be diverted into a productive means to maintain order. In his discussion of Hobbes, Connolly goes even further by suggesting that the potentially dangerous passions simultaneously constitute the very founding force of order.

Connolly rejects the widely accepted view that Hobbes's political theory rests on the assumption of self-interested individuals:

16. *Ibid.*, p. 102.

17. Michael C. Williams, too, argues that Hobbes's aim is to construct "a recognisably liberal politics of toleration and self-creation". However, since Williams seeks to contextualise Hobbes's political theory within the Realist tradition, it will not be further elaborated upon here. See Michael C. Williams, *The Realist Tradition and the Limits of International Relations* (Cambridge: Cambridge University Press, 2005), pp. 35, 38–39.

18. George Kateb, "Hobbes and the Irrationality of Politics", *Political Theory*, Vol. 17, No. 3 (1989), pp. 355–391.

19. *Ibid.*, p. 356.

20. *Ibid.*, p. 383.

21. *Ibid.*, p. 366.

> ... this interpretation exaggerates and misleads. It exaggerates by treating human beings who are to become both self-interested and principled as if they were secure agents of self-interest prior to the education they receive in civil society, and it misleads by pretending that the self-interested individual is the problem when it comes closer to being the solution Hobbes offers for the problem he identifies.[22]

Referring to Hobbes's fundamental conviction that "man is made fit for society not by nature but by education", Connolly suggests that the self-interested individual is not so much the starting assumption as the end product of Hobbesian politics. For Hobbes, the self-interested self is an artifice conducive to a well-ordered society.[23] Accordingly, the main problem of Hobbesian political rule is to convert a being that is initially dominated by his passions into a calculating, prudent and self-interested subject. Crucially, the domain of human passion constitutes not simply a threat to political order but also an important instrument for its reproduction. In Hobbes, Connolly argues, the domestication of life proceeds in part through the deployment of the fear of death, which induces subjects to regulate themselves and to accept external regulations.[24] This fear, Connolly points out, is a passion necessary for the establishment and preservation of an orderly society.[25] Thus, while the production and re-production of political order functions mainly through the containment of passion, the latter, in the form of fear, also plays an important role in this endeavour.

Despite admitting to the productive potential of such disorderly forces, the productivity of its force is dependent on its proper regulation into a movement reproductive of order. The issue of belief, which, according to Connolly, holds together Hobbes's political theory, illustrates this. "It is not a matter of inducing belief for the sake of civil peace", Kateb suggests, "but of regulating belief so that it not become a pretext for civil war."[26] Similarly, Connolly describes Hobbes's method as "part of a political strategy of regulation and control": "it operates to delegitimize irregularity, disruptive thoughts, and actions flowing outside the channel of Hobbesian reason".[27] From both accounts Hobbes emerges as a theorist concerned with movement; a theorist, moreover, who seeks not to arrest movement but to circulate it into a movement (re)productive of order. That this is a continuous effort becomes evident from Connolly's claim that the self-interested individual is not produced once and for all. The self-interested individual, who abides by the rules of orderly society, must be continuously moulded. In fact, Connolly suggests, this individual is partly self-productive. Thus,

> ... because it is a human artifice containing internal resistance, its form must be maintained by the vigilant application of rules and policies; because its creation is essential to individual security, the individual

22. William Connolly, *Political Theory & Modernity* (Ithaca, NY: Cornell University Press, 1993), pp. 26–27.

23. *Ibid.*, p. 29. This point is also made by Williams. See Williams, *op. cit.*, p. 33.

24. *Ibid.*

25. *Ibid.*

26. Kateb, *op. cit.*, p. 368. Williams, too, stresses the importance of belief in this regard, and in Hobbes's political theory more generally. See Williams, *op. cit.*, p. 30.

27. Connolly, *op. cit.*, p. 34.

> participates in the creation and protection of its own self; because its form is crucial to social order, the sovereign power attends to its maintenance.[28]

The Hobbesian state, albeit concerned with the regulation of life and flows of movement through the strict application of rules, does not determine life; it is, rather, according to Connolly, a regulator of conduct. He quotes Hobbes on this point:

> But because all the motions and actions of subjects are never circumscribed by laws, nor can they be, by reason of their variety; it is necessary that there be infinite cases which are neither commanded nor prohibited, but every man may either do or not do them as he lists himself.[29]

Neither Kateb nor Connolly further develops these insights on the significance of the regulation of movement and conduct. Kateb locates Hobbes's liberal credentials in the emancipatory potential of his theory, leaving the regulation of movement out from his main argument. Connolly goes some way towards the identification of movement as a key force by pointing at the significance for Hobbes of the creation of a specific kind of subject(ivity), which must be moulded continuously and is partly self-productive and self-regulative. These aspects, as well as the margin of freedom the state allows the individual and its role as a regulator of conduct, can be understood as liberal aspects of Hobbes's political theory. Connolly does not, however, articulate it as such, nor does he directly link the notions of regulation, freedom, force and movement. It is precisely at the encounter of these aspects—in a play of movements between the regulation of circulation and the productivity/danger of disorderly forces—this paper will go on to demonstrate, that we can identify Hobbes as a liberal thinker.

Natural Science in Movement

Hobbes wrote at a time of revolution and change. Upheaval characterised not only the politics but also the scientific thought of the 17th century. Galileo's scientific experiments, which, through the unification of motion and rest, radically challenged the Aristotelian cosmology, would profoundly transform the understanding of the natural as well as the social world. Hobbes's account of political order is, for instance, an extension of Galileo's mechanical law of motion. His political thought is furthermore influenced by the vitalist William Harvey, who discovered the circulation of the blood. Yet, in several respects Hobbes, as well as Galileo and Harvey, remained Aristotelian in his outlook. The genealogical context of these developments sketched in this section serves to contextualise Hobbes's notion of politics as movement. This section thus demonstrates the ways in which Hobbes's account of political order as based on the principle of movement is closely related to and heavily conditioned by discoveries in the mechanical and natural sciences. This insight paves the way for the evaluation of Hobbesian movement on the level of ontology, which necessitates a re-ordering of the Foucauldian separation between war and circulation.

28. *Ibid.*, p. 30.
29. Quoted in Connolly, *op. cit.*, p. 35.

For Aristotle, movement is limited and finite: change has a definite beginning and end, or telos, its completion, at which point movement will cease and a natural state of rest returns. That is, movement will come to an end once it has arrived at its proper place.[30] Aristotle thereby sets up an opposition between movement and rest, or order and change. Moreover, all motion between objects can be reduced to push and pull.[31] The occurrence of motion requires an external force or agency of movement and ceases as soon as this cause disappears. The cause of motion, according to Aristotle, is its end, the attraction towards a purpose or ultimate goal.[32] Hence, both the initial founding tendency and the end product of a development constitute its nature. This applies to the natural as well as to the social world. The state, as the final stage of the development of human society and the initial tendency towards which humans proceed, is a natural institution, whilst man's social and political qualities are similarly natural.[33]

This account of motion remains firmly entrenched until the dawn of the modern era. Galileo's experiments not only radically transform the notion of movement in mechanics, but equally influence the political thinkers of the time in their reflections on the nature of man and the character of political society. Rather than moving from means to end, from beginning to destination, at which point a natural state of rest would recur, Galileo argues that a body, once set in motion, continues to move along a straight line ad infinitum unless deflected by an outside force. Movement has lost its telos, its fulfilment. Moreover, the distinction between movement and rest has lost its relevance and is replaced by that between movement and acceleration. This new law of motion discovered by Galileo—"any velocity once imparted to a moving body will be rigidly maintained as long as the external causes of acceleration or retardation are removed"[34]—is reflected in Hobbes's idea that human beings can be regarded as matter in motion, who seek only the continuation of motion.[35] The Sovereign is accordingly concerned with the regulation rather than the fixation or impediment of movement.

Significantly, these insights from mechanics come with a new Scientific Method. In order to ascertain its nature, the Aristotelian tradition examines the end of a thing or phenomenon.[36] Aristotle's method of enquiry furthermore consists in breaking up a thing into its simplest elements in order to arrive at first principles.[37] Resolution is key to the Scientific Method, too. To arrive at the nature of a thing, this method holds, one must dissect it into its elementary parts and subsequently recompose it.[38] Hobbes was a proponent of the Method, which, in his

30. Aristotle, *Physics* (Oxford: Oxford University Press, 1996), pp. 186–187; Thomas Spragens, *The Politics of Motion: The World of Thomas Hobbes* (London: Croom Helm, 1973), pp. 56–58.

31. Aristotle, *op. cit.*, pp. 170–171; Martin Schönfeld, *The Philosophy of the Young Kant: The Precritical Project* (Oxford: Oxford University Press, 2000), p. 22.

32. Spragens, *op. cit.*, pp. 57–59.

33. Aristotle, *The Politics* (London: Penguin, 1992), pp. 59–60; Richard Kraut, *Aristotle: Political Philosophy* (Oxford: Oxford University Press, 2002), pp. 59, 247–248.

34. Galileo, *Dialogues Concerning Two New Sciences* (New York: Macmillan, 1914), p. 215.

35. Loralea Michaelis, "Hobbes's Modern Prometheus: A Political Philosophy for an Uncertain Future", *Canadian Journal of Political Science*, Vol. 40, No. 1 (2007), pp. 103–104.

36. Spragens, *op. cit.*, p. 105.

37. Aristotle, *op. cit.*, 9. See also John Watkins, *Hobbes's System of Ideas* (London: Hutchinson, 1973), p. 32.

38. Thomas Hobbes, "De Corpore", in *The Elements of Law, Natural and Politic. Human Nature and The Corpore Politico*, p. 194. See also Spragens, *op. cit.*, pp. 105–117; Watkins, *op. cit.*, pp. 31–32.

view, applied equally to all branches of knowledge. Although this form of enquiry is the basis of geometry, ratiocination is, according to Hobbes, not merely an exercise in numbers. Geometry and physics, the first part of philosophy, provide the method for understanding the other branches of philosophy, for all kinds of philosophy are capable of addition and subtraction. Thus, second, moral philosophy considers the motions of the mind—appetite, aversion, hope, fear, anger, etc. The first beginnings of knowledge are phantasms of sense and imagination; to find the cause of such phantasms one must proceed through the method of resolution and composition.[39] Third, civil philosophy is concerned with the cause and necessity of constituting commonwealths. The method of resolution and composition must be applied to understand politics, for "the principles of the politics consist in the knowledge of these motions from the knowledge of sense and imagination".[40] In other words, knowledge of the state and political affairs proceeds through the resolution of the body politic into its constitutive parts—man's passions and capabilities—and their subsequent recomposition into an ordered state of rational men.[41]

> For everything is best understood by its constitutive causes. For as in a watch, or some such small engine, the matter, figure, and motion of the wheels cannot be well known, except it be taken insunder and viewed in parts; so to make a more curious search into the rights of states and duties of subjects, it is necessary, I say, not to take them insunder, but yet that they be so considered as if they were dissolved; that is, that we rightly understand what the quality of human nature is, in what matters it is, in what not, fit to make up a civil government, and how men must be agreed amongst themselves that intend to grow up into a well-grounded state.[42]

This methodological resolution to discover the human condition consists in the examination of man's mental and physical motions, alone as well as in relation to other men and society. Hobbes dissects the body politic into the passionate nature of man, which, he argues, is the basis of the miserableness and insecurity of life in the state of nature, and recomposes it into an ordered, rational, peaceful sovereign state. The exercise of resolution and recomposition thus appears radically creative: from passionate, irrational beings in a state of insecurity and war this method creates equal bricks with which a stable edifice can be build; rational, homogeneous individuals that fit the mould of an ordered society. However, as will be discussed below, the danger of a disruption of order and a relapse into a state of nature remains the permanent backdrop of the ordered state, for man's motive passions cannot (and must not) be eradicated. The production of order must therefore be followed by its continuous reproduction via circulation.

The state, for Hobbes, rather than a natural development and place, constitutes an artificial construct produced through the mechanical infinity of motion. In opposition to Aristotle, for Hobbes, natural order is geometric only.[43] However,

39. Hobbes, "De Corpore", *op. cit.*, pp. 188, 194, 199–200.

40. *Ibid.*, p. 200.

41. Spragens, *op. cit.*, p. 152.

42. Hobbes, "De Cive", in Bernard Gert (ed.), *Man and Citizen* (Indianapolis and Cambridge: Hackett, 1991), pp. 98–99.

43. Spragens, *op. cit.*, p. 107.

whilst often understood to seek to dominate nature—for nature is disruptive, disorderly and dangerous—Hobbes here rather endeavours to examine nature, and to work with and according to its principles. Although Hobbes refers to the sovereign as "artificial man", and the individual must be shaped into an equal brick to fit the mould of society, here Hobbes suggests the existence of a certain naturalness of man that must be taken into account for this artificial construction. The relationship between this nature and the institution and preservation of (political) order are best understood by considering a second vital influence on Hobbes. Galileo's influence on Hobbes's thinking is often quoted. Hobbes's conceptualisation of the motion of men is described, for instance, as "a human equivalent of the law of inertia".[44] Much less appreciated is the way in which the anatomist William Harvey shaped Hobbes's understanding of both politics and life.[45]

Whilst the Scientific Method allows one to attain universal knowledge in the realms of physics and geometry, as well as moral and civil philosophy, to find the cause of things, Hobbes contends, does not require a method, for all things have one universal cause: motion. Motion itself, Hobbes believes, cannot have any other cause than motion.[46] Motion has become the fundamental principle of nature, order, life, and politics. Hobbes derives this understanding of motion not from the scientific laws of mechanical bodies but from the natural force of the living body. Through the dissection of bodies—another instance of resolution—Harvey discovers the circular motion of the blood in the human body. Whereas conventional wisdom held that blood was produced by the liver (venous blood) and the heart (arterial blood) and subsequently flowed to other parts of the body, where it was consumed, Harvey's anatomical examinations display that the heart was the vital organ that continuously pumped blood around the body. Harvey thereby overthrows Galenic philosophy, which accords primacy to the liver, and returns to the Aristotelian idea that the heart constitutes the primary organ.[47] According to Harvey, the blood would circulate through the body, return to the heart and be recycled into the system. The motion of the blood thus constitutes a continuous, equable, incessant stream, and the heart, as the source of motion, becomes the foundation of life.[48] Dedicating his tract to the Prince, Harvey even likens the heart to the sovereign, who must be understood as the heart and foundation of the life of the state:

> The heart, ready furnished with its proper organs of motion, like a kind of internal creature, existed before the body. The first to be formed, nature willed that it should afterwards fashion, nourish, preserve, complete the entire animal, as its work and dwelling-place: and as the prince in a kingdom, in whose hands lie the chief and highest authority, rules over all, the heart is the source and foundation from which all power is derived, on which all power depends in the animal body.[49]

44. *Ibid.*, p. 177.

45. For an account of Galileo's influence, see, for example, Spragens, *op. cit.*; Harvey's influence is discussed in Watkins, *op. cit.*

46. Hobbes, "De Corpore", *op. cit.*, p. 197.

47. Richard S. Westfall, *The Construction of Modern Science: Mechanisms and Mechanics* (Cambridge: Cambridge University Press, 1977), p. 188.

48. William Harvey, *On the Motion of the Heart and Blood in Animals* (1628), pp. 22–23, 34, 42.

49. Harvey, *op. cit.*, p. 42.

Where Harvey discovers that the function of the heart in the human body is just like the function of the sovereign of a commonwealth, Hobbes argues that the body politic can be understood best by looking at the workings of the human body. For both men these bodies are connected methodologically, for both the human body and the body politic can be understood via the method of resolution and composition. Although Harvey later compromises the importance of the heart and accords a more prominent role to the circulation of the blood, the heart/sovereign remains the source of life.[50] The tension between the importance of, and relations between, the sovereign and the movement of circulation are equally present in Hobbes.

Hobbes distinguishes two types of motion: vital motion and animal motion. Directly referring to Harvey, Hobbes defines vital motion as the motion of the blood that perpetually circulates in the veins and the arteries.[51] For Harvey and Hobbes, vital motion constitutes the force of life, as well as the principle according to which political motion can be understood. Ultimately, therefore, for both men circulation is the prime ordering principle of both life and politics. Voluntary or animal motion is described by Hobbes as the quickening or slackening of vital motion as a result of man's passions. All passions of the mind, Hobbes explains, consist of appetite and aversion; and appetite and aversion are the first endeavours of animal motion.[52] Harvey, too, believes that the motion of the blood is influenced by the affections of the mind.[53] Unlike Hobbes, who is considered a champion of mechanical thought, Harvey is a vitalist, who ascribes spiritual qualities to the blood. Hence, it is interesting to note (albeit beyond the scope of this paper to explore more fully) that the mechanist Hobbes adopts the Harveian term "vital motion".

As the difference between continuous motion and acceleration/retardation, the distinction between vital and voluntary motion is in line with the new Galilean relations of movement: as the distinction between movement and rest has lost its relevance, (continuous) movement becomes opposed to acceleration. The difference between vital and voluntary motion is that between the continuous, equable circulation that ensures and maintains life, and the acceleration and/or retardation of this movement in response to this life's situation in a world. For, according to Hobbes, this motion is ultimately a reaction to the action of external objects.[54] Hobbes does not elaborate upon the interaction between the motion of the human being and that which surrounds him. The full exploration of the relation between man and his environment does not occur until the adoption into biology of Newton's mechanistic concept of the milieu. To better understand Hobbes's conception of movement and circulation in relation to later political thought, let us look beyond Hobbes and briefly return to Foucault's conceptualisation of the birth of "liberal" circulation.

Foucault traces the emergence of the concept and practice of the milieu to a number of interrelated developments, or problems, in the 18th century. These are all problems of circulation and action at a distance. In mechanics, for

50. John Rogers, *The Matter of Revolution: Science, Poetry, and Politics in the Age of Milton* (Ithaca and London: Cornell University Press, 1996), p. 29.

51. Hobbes, "De Corpore", *op. cit.*, p. 226.

52. *Ibid.*, pp. 227–228.

53. Harvey, *op. cit.*, pp. 23, 25, 35.

54. Thomas Hobbes, *Leviathan* (Oxford: Oxford University Press, 1996), p. 35.

example, Newton is occupied with the question of how one body sets another into motion without direct physical contact, which he seeks to resolve via the idea of an ether that permeates all matter and space. At around the same time, the economic and commercial rise of the town prompts its opening up to various forms of movement. City planners become concerned with devising plans that ensure hygiene, trade, surveillance and the connection of the town to external roads. Closure and direct intervention (e.g. the closure of city walls in the evening would stop all traffic) cease to be the preferred, or the only, methods of security; rather, intervention is effected from a distance: "it was a matter of organizing circulation, eliminating its dangerous elements, making a division between good and bad circulation, and maximizing the good circulation by diminishing the bad".[55] In the economic realm of the market, concerning the biological issue of health/disease and in the field of governance, too, notions of circulation and action at a distance arise around the same time in the context of the problem of security.[56]

Put differently, the emergence of the milieu should be understood in relation to the concurrence and/or interaction of several movements: from mechanics to biology; from man to species; from disciplinary power to mechanisms of security; from contiguous circulation to laissez-faire circulation or action at a distance; from contiguous motion to the milieu. Foucault suggests that, whereas it arises first as a practice, the emergence of the concept (not yet the term) milieu can be traced to Newton.[57] Newton's concern with the possibility of action at a distance connects the mechanical theorists' new scientific ideas of motion—not only Galileo's but also Descartes'—and the political thought on which it bore influence, with its subsequent transformation, which is ultimately expressed—in biology, in politics, and in other domains—in the relation between man and milieu.

Although the possibility of action at a distance via the medium of ether is first elaborated by Newton, his conceptualisation remains ambiguous on this point. Initially he understands the ether to be a fluid of minute particles that pervade all space, whereas he later poses the idea of forces that act between bodies.[58] This latter idea of forces of attraction between bodies allows the possibility of action at a distance. Regardless of this ambiguity concerning the existence of an ether and its precise character, Newton's significance in the context of the milieu lies in posing a medium of action. For Galileo and Descartes there is only matter and motion. Newton discovers that force is not a property of bodies, but acts in between.[59] As pointed out by Canguilhem, upon which Foucault bases his analysis, it is this idea of a medium of action—a medium that transmits action over a distance—that is subsequently expressed through the term milieu in biology.[60] In economics, the notion laissez-faire similarly expresses this idea: intervention is required and desired only from a distance.

55. Foucault, *Security, Territory, Population*, *op. cit.*, p. 18.

56. *Ibid.*, especially lecture three.

57. *Ibid.*, p. 20. Foucault relies on Canguilhem's elaboration of the milieu: Georges Canguilhem, "The Living and its Milieu", *Grey Room 03* (Spring 2001), pp. 7–31 (p. 8).

58. It should be noted that Descartes had already put forward a theory of the ether. However, according to Descartes, the universe is a plenum, and all space, including the ether, is matter. His theory thus remains premised on the impossibility of action at a distance: all transmission of motion must occur via impact. See Westfall, *op. cit.*, pp. 33–34, 119, 120, 121.

59. Schönfeld, *op. cit.*, p. 67; Westfall, *op. cit.*, pp. 140–143, 157.

60. Canguilhem, *op. cit.*, pp. 1–2.

And, as Foucault further explains, the practice of security through which liberal government governs works via ensuring, fostering and managing circulations from a distance.[61]

Although Hobbes's analysis pre-dates the developments that led to the conceptualisation and practice of the milieu, his thought can be seen as a precursor to these developments. First, Hobbes conceptualises the fundamentality of circular motion. Although Aristotle ascribes a circular motion to the rise and fall of political systems, this movement lacks the centrality it gains with Harvey and Hobbes. Whereas Harvey later downgrades its role, for Hobbes the heart/ sovereign remains the centre of circulation; and, for both men, circulation constitutes the centre of life. In other words, circulation becomes ontological. Second, although his account lacks a conception of environment, there is nevertheless a first inkling that motion changes in response to environmental factors. Yet, writing prior to Newton, the rise of the town, and the development of biology, for Hobbes circulation lacks a medium of action and remains couched in Aristotelian terms, as a movement of push and pull:

> I have shown that no motion is generated but by a body contiguous and moved: from whence it is manifest, that the immediate cause of sense or perception consists in this, that the first organ of sense is touched and pressed. For when the uttermost part of the organ is pressed, it no sooner yields, but the part next within it is pressed also; and, in this manner, the pressure or motion is propagated through all the parts of the organ to the innermost. And thus also the pressure of the uttermost parts proceeds from the pressure of some more remote body, and so continually, till we come to that from which, as from its fountain, we derive the phantasm or idea that is made in us by our sense.[62]

The consequences of the transformation from contiguous motion to action at a distance are significant: the notion of the milieu gains relevance not only in mechanics and biology but also in the political, economic and social domains, as is very well illustrated in the first three lectures of Foucault's *Security, Territory, Population*. In the context of this paper, most important to note is that whilst circulation continues to function as the main principle of order(ing), it is no longer on the basis of contiguity in order to control directly but rather as a matter of action at a distance; laissez-faire. If movement is ontological, then a transformation of its relationality implies a change in the relations of life. The question is, therefore, to what extent and in which ways the conception of life, as well as the relation between politics and life (for politics is movement) alters with the emergence of the milieu. In addition, the extent to which (some notion of) non-contiguous motion might have already been present prior to the milieu, as well as the continued conception and practice of contiguity after the emergence of the milieu, requires further examination.

The relation between life and movement drawn by Hobbes and Harvey are nevertheless of great significance as a first conceptualisation of a politics of life that moves beyond strictly mechanical movement. For mechanists such as

61. Foucault, *Security, Territory, Population*, *op. cit.*, p. 29.
62. Hobbes, "De Corpore", *op. cit.*, p. 214.

Descartes, the same laws of motion apply to all matter. Jacob ascribes this view to Hobbes, too, whom he identifies as one of the main representatives of the mechanical tradition:

> According to Hobbes, an animal could be considered either as a machine or as automaton whose limbs twitch like those of a man endowed with artificial life. That was not a metaphor, a comparison or an analogy. It was an exact identification. Stars, stones and living beings, all bodies were subject to the same laws of movement.[63]

Although Hobbes does refer to animals in these terms, and to sense in terms of "mutual action and reaction", he *also* characterises them through the notions of vital and voluntary motion. Jacob does not overlook Harvey's influence on Hobbes as much as classify both Hobbes and Harvey as mechanists. Yet the concept of vital motion would seem a peculiar one for an exclusively mechanistic thinker. Moreover, the differentiation between continuous circular motion and voluntary or animal motion as a reaction that seems not merely automatic or mechanical but voluntary suggests that, contra Jacob's claim, there is no simple identity between nature and machine in Hobbes. It is through the distinction between vital and voluntary motion that Hobbes distinguishes between the living and the non-living.

Voluntary motion can thus be said to mark the distinction between inert matter and living things. Whereas vital motion could still be likened to Galilean inertial motion, voluntary motion is more complex than its simple perturbation according to mechanical laws. Only organisms are capable of sense perception and the actions and reactions this entails. Even though one could conceive the transmission of motion from one organ to the next described by Hobbes in mechanical terms, the passions and desires—the will or endeavour—involved in the process could hardly be accorded to a stone or a star. Despite his mechanistic conceptions of order(ing), we find in Hobbes *also* an account of the distinction between living and non-living things, drawn on the basis of movement. Significantly, in the 18th century, when the difference between living beings and non-living things becomes more clearly articulated and more thoroughly conceptualised, it comes to be drawn not in terms of movement but of organisation.

As this section has sketched, movement and circulation did not emerge as principles of order(ing) with the rise of 18th-century liberalism. A conception of politics as movement via (the securitisation of) life, which would come to define a specific liberal mode of governance, was already prefigured in Hobbes's thought. The mode of circulation, and concomitant conceptions of order, politics and life have, however, undergone change. It is with Hobbes, via Harvey, that circulation becomes the central ordering principle of life; that circulation becomes ontological. Hobbes thus adopts Harvey's discovery of the circulation of the blood as the vital foundation of his political thought. Despite being insightful, on the one hand, regarding the relations of politics and war in Hobbes and, on the other, regarding the relations between liberalism and circulation, Foucault overlooks both the fundamental significance of circulation in Hobbes's thought

63. François Jacob, *The Logic of Life: A History of Heredity* (Princeton: Princeton University Press, 1993), p. 33.

and its ontological status. Rather than the opposite figure of liberal conceptions of governance and order, as Foucault's analysis suggests, Hobbes can be understood as a founding liberal thinker.

(Re)productive Movement and Political Order

The genealogical sketch of the development of the concept of movement in the natural sciences in the previous section has located Hobbes's account of movement in relation to Galileo and Harvey. The second section of this paper has furthermore situated this understanding genealogically between the Aristotelian and Newtonian notions of movement. Finally, it was suggested that circulation as a principle of ordering both politics and life did not emerge in the 18th century but was already central in Hobbes. This section will more thoroughly explore the centrality of movement in Hobbes's ontology. More precisely, we will suggest, contra Foucault, that war and circulation do have a presence in Hobbes. Moreover, and significantly, both war and circulation must be understood (a) in terms of movement and (b) at the level of ontology. In conclusion, this section will argue that Hobbes's account of political order rests on two forms of movement, or, rather, a double movement: productive and reproductive movement. The theory of the contract, which Foucault problematically takes as his interpretative point of departure, is already part of the logic of (re)productive movement and must be seen in this context.

What is peculiar about Hobbes's use of the concept of the state of nature is that he employs it in two different ways. First, he employs it in an ontological sense in order to capture a human condition without epistemological or political foundations. In this sense, the concept of the state of nature contains a philosophical account of the human condition rather than merely a historical description of the situation that men found themselves in prior to the emergence of state sovereignty. As both Connolly and Michael C. Williams point out, ontologically, the state of nature captures the uncertainty and the fundamental indeterminacy of life that Hobbes's nominalist inclination would have implied.[64] The nominalist tradition—driven by the theological crisis caused by the difficulty of explaining the persistence of evil in the world—removes God from the mundane world in order to save his omnipotence.[65] The consequence of this move—a first step towards the secularisation of the world—is that the human condition loses its divine foundations and becomes radically contingent and uncertain.[66] Hobbes's nominalism finds expression in his concept of the state of nature. The disorder that Hobbes sought to illustrate here is epistemological and political in nature. Language, truth, morality and political order are nothing but arbitrary constructions produced in the hearts and minds of men.

The problem of war is thus, for Hobbes, a problem of diverse and competing rationalities and claims to political authority; the problem is the absence of a definitive judgement on what life and political order should be like. Hobbes's philosophico-political imagery corresponds to a field of forces, the constituent elements of which are different and competing wills to truth/knowledge,

64. Connolly, *op. cit.*, p. 19; Williams, *op. cit.*, pp. 22–25.
65. Connolly, *op. cit.*, p. 19.
66. *Ibid.*, pp. 18–20.

which, in turn, inspire different and competing wills to power/authority. Hobbes's security dilemma is therefore much closer in nature to what is often referred to as the paradox of modernity than to the rationalist formulation usually forwarded by theorists of international relations. That is, the problem of war is, for Hobbes, not so much a problem of the competing interests of pre-constituted rational subjects but a problem of immanence and transcendence, identity and difference. It is the multiplicity of wills to knowledge and power and the multiple and mutually exclusive pursuits of security emanating from the immanent condition of mankind that produce the problem of war. This field of warlike forces presents a different mode of movement than the vital circulation, which sustains sovereign order. It is, rather, a complexity of disorderly movements, or a field of forces in disorderly movement, producing all manner of encounters and clashes. Interestingly, this idea of a field of forces comes close to Foucault's own notion of power conceptualised as a field of force relations. Due to the lack of an overarching authority—a sovereign will that demarcates and bounds off the field—life, order, truth, etc. remain indeterminate and uncertain. This ontological state should be understood as a movement rather than a historical state—that is, it can be understood productively only in relation to the imposition and reproduction of order via circulation; on its own, it is unsustainable. A consideration of the second mode of the state of nature will allow a further exploration of this relation between the movement of war and that of circulation.

The second meaning of the state of nature is conferred on it in relation to the logic of the contract. As has been pointed out by a number of scholars, the contract is best understood as ideological myth. Thus, ideologically, the state of nature helps in presenting an account of the origin of order whose function is to erase the arbitrariness and violence that necessarily accompanies the attempt to impose order on the kind of immanent condition of life that Hobbes recognised. In a sense, then, the state of nature is interesting both in relation to the original problem that Hobbes encountered as well as in relation to the solution that he formulated; it tells us what he believed the world to be like and also what he believed people should believe it to be like in order for political organisation to be effective. The problem with Foucault's reading of Hobbes is that he considers only one part of the story. It is true that Hobbes sought to conceal the foundational violence that necessarily accompanies the emergence of political authority. But there is more to Hobbes than this, and had Foucault considered the other dimension of Hobbes he would have found it far more difficult to establish a clear opposition between Hobbes and himself.

Hobbes's aim is to create a "body politic", an "artificial animal" whose principle of organisation is, as has been shown in the last section, that of circulation. Vital motion or circulation is the organic principle of organisation on the basis of which political order could function. To create a body politic, therefore, the voluntary motions of individuals have to be made vital or life-sustaining to political order. The function of the heart is to be represented by the power of the sovereign who gives "life and motion to the whole body".[67] Out of the messy field of forces created by different pursuits of security, Hobbes seeks to create an order of circulation in which individual parts work for the security of the whole. The body politic strives not towards a telos or towards the actualisation of the potential

67. Hobbes, *Leviathan, op. cit.*, p. 9.

but towards persistence or security. For Hobbes, security is defined on the basis of the distinction or the fundamental antagonism between the forces of reason and the forces of passion. This is the war that underlies Hobbesian political order; it is a war that can never be won once and for all, yet a war that determines what is and what is not to be included in the realm of political order.

War and circulation meet where the immanent field of forces, in which relations remain open and undetermined, is demarcated and bounded; where contingency and indeterminacy become circulated into definite and sovereign claims to order, life, subjectivity, method, knowledge, etc. Thus, via the principle of circulation (the problem of) war is striated into a rationalised field of orderly interactions. Despite this movement towards a field of circulation, the war can never be won, for disruptive forces remain immanent within this field. Whilst circulation becomes the principle of order(ing), its very force is derived from the mo(ve)ment of war. Circulation gains its force from the passionate will to live that has been vitalised into circular movement, yet the disruptive quality of which remains dormant. Consequently, the reproduction of circulation constitutes a continuous never-ending effort: the state of order must be reproduced ad infinitum to prevent a return to the state or problem of war, even though the mo(ve)ment that is thereby obfuscated constitutes the very possibility of order and circulation. The meeting of war and circulation or, put differently, the encounter of productive and reproductive movements, thus constitutes the political mo(ve)ment that both creates the possibility of social relations—relations of movement, politics, life—and that determines what these relations can be and may become within the particular order constituted.

Hobbes's elaboration on the cause of motion exemplifies the ambiguity of the nature of this encounter:

> But the causes of universal things (of those, at least, that have any cause) are manifest of themselves, or ... known to nature; so that they need no method at all; for they have but one universal cause, which is motion. For the variety of all figures arises out of the variety of those motions by which they are made; and motion cannot be understood to have any other cause besides motion ...[68]

This idea that there is only movement can be understood to mean that there is merely one mode of continuous movement: circulation. Indeed, Hobbes seeks to obfuscate the violent and disruptive origins of order via the introduction of a number of circularities: war is veiled by presenting the foundations of order as a consensual, rational practice, whilst reason and consensual practices are possible only after order has been founded. Despite his insistence on the use of scientific method, Hobbes contends that no method is required to find the cause of movement. Such enquiry is thus superfluous, as is the effort to unearth the origins of political order: there is only circulation.

This paper has demonstrated what occurs when one *does* explore the origins of Hobbesian political order: whilst there is only movement, there is more than circulation. Rather than the mere, perpetual continuity of circulation the foundations of this movement lie in a break, a disruptive mo(ve)ment. There is only movement,

68. *Idem*, "De Corpore", *op. cit.*, p. 197.

yet there are two modes of movement: productive and reproductive. The productive, disorderly field of forces underlies its circular order(ing) that marks out, establishes, and continuously reproduces the limits of what political movement can be. The relation between these two movements is, however, more complex than this, for whilst reproductive movement is possible only on the basis of its production, and whilst reproduction is continuously fed by the rendering "vital" of animal motions, the productive movement only gains its productive quality through its subsequent circulation. Hence, political movement remains a play of forces in interaction even though, at the surface, one perceives only—and Hobbes desires one perceives only—the ordered field of circulation: the order of the sovereign state.

If the principle of organisation of Hobbes's political order is circulation, then to read Hobbesian political order in terms of juridical order, which Foucault does, is, if not entirely false, then at least partial and somewhat misleading in so far as this reading ignores the fact that the law serves not as a the principle of order for Hobbes but as a tool that works tactically in support of a principle of order derived not from the legal but from the domain of natural philosophy. This tactical function of the law, which Foucault associates exclusively with biopolitics, becomes apparent when Hobbes asserts that "... laws were not invented to take away, but to direct men's actions; even as nature ordained the banks, not to stay, but to guide the course of the stream."[69] The law as well as the disciplinary regime that Hobbes points at serve to regulate the voluntary movements of individual bodies so as to make these conducive to the vital motion of the body politic, the latter being defined in terms of security.

The point is this: if the principle of order is not the law but circulation then political order is constituted by the permanent production of movement and not by a single movement from a lawless condition to the state of law. The sovereign is then not simply the one who sees to it that individual movements conform to the law as an end in itself; she is a producer of movement. Individuals do not circulate in the body politic in the sense of being fully constituted and moving within a prearranged political space; they circulate *as* the body politic. The movement that we should look at if we want to rediscover Hobbes as a site for thinking critically about politics is that of circulation, and not the movement expressed in terms of a transition from war to peace. Political order, for Hobbes, was not a condition that could be achieved once and for all through some form of transition, and it did not rest on a static form of state sovereignty. Instead, Hobbesian political order amounts to the permanent production of a striated, circular movement; sovereign power is the product and the producer of a regulated, circular, movement. Circulation reproduces the form of life deemed suitable for inclusion into political order. Political life is in need of being permanently (re)produced.

Thus, we find in Hobbes two modes of movement, which together constitute and perpetuate political order: productive and reproductive movement. Since these are not separate movements but rather constitute a double movement, one can conceptualise Hobbesian political order as (re)productive movement. Whilst the political mo(ve)ment is the first encounter of these movements, their continuous play of interaction—clashing, merging, (re)producing, (de)constructing—constitutes the politics of order(ing), i.e. the management of circulation. In short,

69. *Idem*, "De Cive", *op. cit.*, p. 268.

Hobbes's account of political order is marked by a continuous play of (re)productive movement(s).

Conclusion

This paper has demonstrated that the political theory of Thomas Hobbes offers more than the depoliticising image of the transition from the state of nature to civil society. Underneath this motif, we have rendered visible an understanding of political order as (re)productive movement, through which Hobbes's account can be understood to be of profound significance for liberal political thought. The paper set out with a discussion of Foucault's conceptualisation of power in the context of his reading of Hobbes, which revealed the limitation of Foucault's Hobbes to the theory of the contract. While we endorse Foucault's claim that Hobbes provided liberalism with the apolitical motif that continues to define its problematic self-image, we would reject the claim that this is all there is to Hobbes. Therefore, Foucault's distinction between a Hobbesian conception of power in terms of peace and law on the one hand, and his own conception of power in terms of war and circulation on the other, is misleading. Departing from Foucault's analysis, for whom Hobbes remains the theorist of the static image of peace, sovereignty and the law, we have argued that Hobbes is also the theorist of movement and war.

Movement—both as war and as circulation—is central to Hobbes's thought. Understood at the level of ontology, these two movements are inextricably linked—circulation, one could say, is a continuation of war by other means, and vice versa. Approached via the notion of a play of (re)productive movement(s), Hobbes's thought offers an understanding of political movement that can still be useful today for developing a critical perspective on liberal regimes of power. What can Hobbes, thus understood, tell us about International Relations? We suggest three things. First, rather than capturing a political space defined by the logic of anarchy, Hobbes's theory of power is suggestive of a political space defined by the logic of circulation. In this sense, Hobbes's value for International Relations lies not so much in the definition of a realist image of the international as in the articulation of a distinctly liberal problematisation of political space. The liberal political rationality of global security is increasingly preoccupied with the problem of circulation, that is, with differentiating between good and bad circulation and with maximising the good while minimising the bad. To be sure, the security of global circulation is related to complex global interdependencies and dynamics that Hobbes could not have conceptualised. Nevertheless, Hobbes can be said to have first articulated the principle upon which the security of global liberal governance operates today.

Second, Hobbes's political theory reveals that the occurrence of a principle of circulation does not indicate a move beyond sovereign power; it is, on the contrary, premised upon it. Moreover, Hobbes provides us with an idea of how sovereign power is related to and works within this space of circulation. It is through Hobbes and through the logic of (re)productive movement that we can understand, for example, the way in which the liberal, biopolitical space of circulation is bound to a principle of formation through which a certain form of life is produced and reproduced. What we can understand, in other words, is that what is at stake in the liberal space of circulation is, in the first instance, the security of circulation. It is therefore through this logic of (re)productive movement that we can also understand that circulation is always bound to the logic of sovereign power, which enacts the break that separates order from disorder and one form of life from another.

Finally, the two movements at play in Hobbes's account of political life, being inextricably linked and in perpetual interaction, continuously (re)produce effects of both war and peace, order and disorder, power and resistance, which suggests that the fine line between power and resistance that is central to Foucault's political thought is—in a different way—present in Hobbes, too. Despite his attempts to veil the disorderly forces at work, in Hobbes we find recognition not only of the idea that power and resistance are inextricable but also that resistance, in a way, precedes power. While we must recognise that the problematisation of circulation, the workings of sovereign power, and the possibilities of resistance have all undergone significant transformations since the 17th century, it is fair to suggest that Hobbes was the first to theorise all these elements of modern (liberal) politics together.

Taking Foucault beyond Foucault: Inter-state Governmentality in Early Modern Europe

HALVARD LEIRA*

The analysis of governmentality has had a profound impact on the study of liberal, domestic societies over the last two decades, and the conceptual framework has been applied successfully to current global affairs. In this article one possible way of expanding the timeframe and the scope of governmentality studies is explored. Through an immanent critique of Foucault's own comments on the co-constitutive development of states and a state system in early modern Europe, it is argued that a governmentality perspective can in fact add to our understanding of inter-state relations in early modern Europe, and thus also to our understanding of our own time. Carrying out such analyses implies taking the Foucauldian framework beyond Foucault, as his own brief comments on inter-state relations fail to adhere to his own methodological precept of historicising seemingly evidentiary practices.

> [I]t is the genealogy of problems that concerns me. Why a problem and why such a kind of problem, why a certain way of problematizing appears at a given point in time.[1]

Introduction

During the last two decades, a growing literature has approached power through the concept of "governmentality", coined by Michel Foucault in the late 1970s. Most of the studies have dealt with domestic societies, and have focused on (relatively) current affairs. While there is a growing body of International Relations (IR) studies applying a governmentality perspective to current affairs, historical work has largely been confined to the inside of the modern state.

The publication of *Security, Territory, Population*,[2] the full set of lectures where Foucault developed the concept of governmentality for the first time, however,

*Thanks are due to the two anonymous reviewers for insightful and stimulating comments, and Nicholas Kiersey for pulling this issue together and keeping spirits up. Full responsibility for any misunderstandings or errors remains with the author.

1. Michel Foucault, "What Our Present Is", in Sylvère Lothringer (ed.), *The Politics of Truth* (Los Angeles: Semiotext(e), [1981] 2007), p. 141.

2. Michel Foucault, *Security, Territory, Population: Lectures at the Collège de France 1977–1978* (Houndmills: Palgrave, 2007).

makes possible a much wider and more historically grounded approach. In these lectures, Foucault not only expounded on the processes and fundamentals involved in the study of government, he explicitly connected domestic and external developments. Like many IR thinkers of recent decades, he seems to have seen the state and the inter-state system as mutually constitutive. It follows from this that ideas about how the state should act in the inter-state system are closely intertwined with ideas about what the state is like. Thus, a governmentality perspective should in principle be applicable to inter-state affairs from the same time as Foucault applied it to domestic affairs—around 1600. In this article I explore whether it indeed makes sense to apply a governmentality perspective to inter-state relations in early modern Europe, what such an application might look like and what can be gained from using it. With a nod to the epigraph, it will be argued here that the emergence and development of inter-state governmentality could profitably be studied through the continuous problematisations of sovereignty, diplomacy and foreign policy and their seeming permanence—how did sovereignty, diplomacy and foreign policy arise as problems when they did and in the way they did, and how have they retained at least the veneer of permanence?

The first part of this article sets the stage by establishing the conceptual baseline of how governmentality and related terms will be understood here and by presenting the place for inter-state governmentality within Foucault's thought. The main part of the article then presents the case for inter-state governmentality. The case is made through an immanent critique of Foucault's own comments on the international. His comments on the role of government in inter-state relations are pithy and thought provoking, but his reliance on a relatively select set of texts implies, on the one hand, that he has missed out on important nuances, and on the other that he underestimates the changes that did indeed take place in inter-state relations between *c.*1600 and *c.*1850. The conclusion will then hint at how Foucauldian insights can be used to push the historical study of inter-state governmentality beyond Foucault.

Governmentality—Analytics and History

Power

The concept of governmentality must necessarily be understood in the wider context of Foucault's ideas about power, and thus a brief note on this understanding must be made. In general, Foucault argued for a move from the questions of "what?" and "why?" to "how?"—a move beyond both the traditional and the radical conceptions of power,[3] and on to what has been called the "fourth face" of power,[4] or productive power.[5] In an oft-cited passage, Foucault criticises traditional power analyses for ignoring "the productive aspect of power", and comments that

> What makes power hold good, what makes it accepted, is simply the fact that it doesn't only weigh on us a force that says no; it also traverses and

3. Stephen Lukes, *Power: A Radical View* (Houndmills: Macmillan, 1974).
4. Peter Digeser, "The Fourth Face of Power", *Journal of Politics*, Vol. 54, No. 4 (1992), pp. 977–1007.
5. Michael Barnett and Raymond Duvall, "Power in International Politics", *International Organization*, Vol. 59, No. 1 (2005), pp. 39–75.

> produces things, it induces pleasure, forms knowledge, produces discourse. It needs to be considered as a productive network that runs through the whole social body, much more than as a negative instance whose function is repression.[6]

This is a take on power that differs significantly from the proprietary notion of power as capacity, and that shifts attention away from actors wielding sovereign power to subjects whom power works through and the processes and relations that thus arise. Such power relations are notable for being "a mode of action that does not act directly and immediately upon others. Instead it acts upon their actions: an action upon an action, on possible or actual future or present actions."[7]

Asking how power is exercised "concretely, and in detail" leads Foucault to specify that one should look at the "specificity, techniques and tactics" of such exercise rather than focusing on the will to domination and power-strategies.[8] The ideal is to get to the micro-practices of power, to study it where it works in detail, on and through specific subjects, both through immaterial and material processes.[9]

Government and Governmentality

Foucault mainly studied local sites and the microphysics of power, but in the late 1970s he took a new interest in the state and power. When looking back on his previous work in 1978, he stressed how he had attempted:

> to free relations of power from the institution, in order to analyze them from the point of view of technologies; to distinguish them also from the function, so as to take them up within a strategic analysis; and to detach them from the privilege of the object, so as to resituate them

6. Michel Foucault, "Truth and Power", in James D. Faubion (ed.), *Power: Essential Works of Foucault*, Vol. 3 (London: Penguin, [1977] 2000), p. 120.

7. Michel Foucault, "The Subject and Power", in James D. Faubion (ed.), *Power: Essential Works of Foucault*, Vol. 3 (London: Penguin, 1982 2000), p. 340.

8. Foucault, "Truth and Power", *op. cit.*, p. 117. Cf. also the methodological precaution that "rather than orienting our research into power toward the juridical edifice of sovereignty, State apparatuses, and the ideologies that accompany them, I think we should orient our analyses of power toward material operations, forms of subjugation, and the connections among and the uses made of the local systems of subjugation on the one hand, and apparatuses of knowledge on the other. In short, we have to abandon the model of Leviathan [...] We have to study it [power] by beginning with the techniques and tactics of domination" (Michel Foucault, *"Society Must be Defended": Lectures at the Collège de France 1975–1976* (New York: Picador, 2003), p. 34). It should, however, be noted that a common reading of Foucault is that his later work on government sought to transcend the dichotomy of sovereignty and domination, seeing power as neither the result of the free play of sovereign wills nor as repressive domination, but instead as productive; Mitchell Dean, *Governmentality: Power and Rule in Modern Society* (London: Sage, 1999), p. 25. A gradual move from battle and domination to government can be perceived by comparing *"Society Must be Defended"* with *Security, Territory, Population*. Cf. Michel Sennelart, "Course Context", in Foucault, *Security, Territory, Population*, *op. cit.*, pp. 369–371.

9. As noted, and criticised, by Jan Selby, there has been a tendency in Foucauldian-inspired IR to focus on meta-theoretical critiques and text alone, rather than the micro-practices and the importance of materiality and practices, in addition to text, in the study of discourse; Jan Selby, "Engaging Foucault: Discourse, Liberal Governance, and the Limits of Foucauldian IR", *International Relations*, Vol. 21, No. 3 (2007), pp. 324–345.

> within the perspective of constitution of fields, domains, and object of knowledge. If this triple movement of a shift to the outside was tried with regard to the disciplines, I would now like to explore this possibility with regard to the state [. . .] Can we talk of something like a "governmentality" that would be to the state what techniques of segregation were to psychiatry, what techniques of discipline were to the penal system and what biopolitics was to medical institutions?[10]

With his lecture series of 1977/78, Foucault thus brought a new concept into his theoretical vocabulary—governmentality, and along with it the notion of governmental power. Both concepts were developed through a reading of long historical changes in modes of power taking place from around 1600 to the late 20th century.

It should be noted right away that neither Foucault nor his commentators have been consistent in the conceptual usage, thus it is necessary to establish the understanding that informs this article. As Foucault sees it, "the general problem of 'government' suddenly breaks out in the sixteenth century with respect to many different problems at the same time and in completely different aspects".[11] Government could relate to oneself, to souls and conduct, to children and to the state; "How to govern oneself, how to be governed, how to govern others, by whom people will accept being governed, how to become the best possible governor—all these problems, in their multiplicity and intensity, seem to me to be characteristic of the sixteenth century."[12] In general, government is about directing conduct: "To govern, in this sense, is to structure the possible field of action of other people."[13] The aim of government is, through the "conduct of conducts",[14] "to affect the actions of individuals by working on their conduct—that is, on the ways in which they regulate their own behaviour".[15] Government is thus not only, and not even primarily, about affecting behaviour directly, but about indirectly affecting how people govern themselves.[16] To Foucault, government thus emerges as a power relation between domination (signifying relationships where the subordinate has little room for action) and the unstable and reversible strategic power relations between sovereign actors.[17] Government signifies a power form with relative stability, but less subordination than under domination.

10. Foucault, *Security, Territory, Population*, *op. cit.*, pp. 118, 120.

11. *Ibid.*, pp. 88, 231.

12. *Ibid.*, p. 88.

13. Foucault, "The Subject and Power", *op. cit.*, p. 341.

14. *Ibid.*

15. Barry Hindess, *Discourses of Power: From Hobbes to Foucault* (Oxford: Blackwell, 1996), p. 97. In a brief throwaway, Foucault describes pastoral power as "government of individuals by their own verity"; Michel Foucault, "Omnes et Singulatim", in James D. Faubion (ed.), *Power: Essential Works of Foucault*, Vol. 3 (London: Penguin, [1979] 2000), p. 312.

16. Michel Foucault, "The Ethics of the Concern of the Self as a Practice of Freedom", in Paul Rabinow (ed.), *Ethics: Essential Works of Foucault*, Vol. 2 (London: Penguin, [1984] 1997), p. 299. As Dean defines it, "Government is any more or less calculated and rational activity, undertaken by a multiplicity of authorities and agencies, employing a variety of techniques and forms of knowledge, that seeks to shape conduct by working through our desires, aspirations, interests and beliefs, for definite but shifting ends and with a diverse set of relatively unpredictable consequences, effects and outcomes" (Dean, *op. cit.*, p. 2).

17. These are, as noted, types or levels of power relations. They are, at least in principle, possible in all forms of human intercourse, and as such are not reducible to the three broad forms of power

For government to make any sense at the state level there had to be both an apparatus that could govern and an interest in governing the subjects rather than just lording over them and perpetuating rule. Government thus, on the one hand, hinged on the phenomenon that "this something, the state, really began to enter into reflected practice",[18] or on the "self-consciousness of government",[19] with political science as it was developing from *c*.1580 to *c*.1650. On the other hand, government came about as a desire grew to conduct the population. On both counts it is immediately clear how government relates to knowledge; the state's knowledge of itself and the state's knowledge of its population: "The modern state is born, I think, when governmentality became a calculated and reflected practice".[20]

It is important to note, however, that government does not replace other forms of power; rather, both sovereignty and discipline are rearticulated and actualised:

> So we should not see things as the replacement of a society of sovereignty by a society of discipline, and then of a society, say, of government. In fact we have a triangle: sovereignty, discipline and governmental management, which has population as its main target and apparatuses of security as its essential mechanism.[21]

Likewise, Foucault stresses that many of the pieces of the state apparatus, such as the army, taxation and systems of justice, were not new to the 16th century but that it was only with the emergence of self-reflection that something like the state could be said to emerge: "The state is therefore a schema of intelligibility for a whole set of already established institutions, a whole set of given realities."[22] It follows implicitly that it makes little sense to be discussing the state as a thing or an entity: "The state is practice. The state is inseparable from the set of practices by which the state actually became a way of governing, a way of doing things, and a way too of relating to government."[23] Taking the logic one step further, Foucault raises the question of whether it is possible to situate the state "as a fundamental political issue [. . .] within a more general history of governmentality, or, if you like, in the field of practices of power".[24] The state then emerges as "nothing more than way of governing [. . .] nothing more than a type of governmentality".[25]

As Dean notes, governmentality for Foucault denotes, on one hand, a general phenomenon, dealing with "how we think about governing, with the different

(sovereignty, discipline and governmental management), on which more below, which are more closely related to polity-centred power and the techniques or mechanisms of power.

18. Foucault, *Security, Territory, Population*, *op. cit.*, p. 247.

19. Sennellart, *op. cit.*, p. 387.

20. Foucault, *Security, Territory, Population*, *op. cit.*, p. 165.

21. *Ibid.*, pp. 107–108. The different rationalities do not correspond directly to any existing system of government. They are the ideal ways of governing that can in principle be described as the underlying principles for ordering knowledge, but they are necessarily different from the actual system of government. Cf. *ibid.*, p. 8.

22. *Ibid.*, p. 286.

23. *Ibid.*, p. 277.

24. *Ibid.*, p. 247.

25. *Ibid.*, p. 248. "The state is not a cold monster; it is the correlative of a particular way of governing" (Michel Foucault, *The Birth of Biopolitics: Lectures at the Collège de France 1978–1979* (Houndmills: Palgrave, 2008), p. 6).

mentalities of government", and with "the art of government", and as such it is closely connected to both explicit and tacit knowledge.[26] On the other hand, governmentality also denotes the historic emergence of one specific way of thinking about and exercising power, starting in 16th-century Europe, on which more below.

The relationship between government and governmentality is not entirely clear-cut in Foucault's own usage, Sennelart even commenting that they seem to merge from the 1979 lectures onwards.[27] At the outset, governmentality was seen to indicate three things.[28] First:

> the ensemble formed by institutions, procedures, analyses and reflections, calculations, and tactics that allow the exercise of this very specific, albeit very complex, power that has the population as its target, political economy as its major form of knowledge, and apparatuses of security as its essential technical instrument.

Governmentality could thus be understood as the many-faceted conditions that enable governmental power to be exercised. Second:

> the tendency, the line of force, that for a long time, and throughout the West, has constantly led towards the pre-eminence over all other types of power—sovereignty, discipline, and so on—of the type of power that we can call "government" and which has led to the development of a series of specific governmental apparatuses on the one hand, and, on the other to the development of a series of knowledges.

Governmentality thus also has a dynamic to it, furthering government and developing its auxiliary apparatuses and knowledges. Third: "The process, or rather, the result of the process by which the state of justice of the Middle Ages became the administrative state in the fifteenth and sixteenth centuries and was gradually 'governmentalized'". Finally, governmentality is here again seen as dynamic, as a key part of the emergence of the modern state.

Although the term became less specific with usage, it seems fair to claim that governmentality in its general form denotes the wider field of knowledge and strategy in which governmental power and practices are established and pursued.[29] The most fruitful, and indeed, Foucauldian, approach seems to be to follow Dean in stressing "analytics of government" rather than specifying

26. Dean, *op. cit.*, pp. 16–19. That there is a relationship between governmentality and mentality seems obvious, even though Sennelart stresses that "the word 'governmentality' could not result from the contraction of 'government' and 'mentality', 'governmentality' deriving from 'governmental' like 'musicality' from 'musical' or 'spatiality' from 'spatial', and designating, according to the circumstances, the strategic field of relations of power or the specific characteristics of the activity of government" (Sennelart, *op. cit.*, p. 399, n. 126).

27. Sennelart, *op. cit.*, p. 388.

28. Foucault, *Security, Territory, Population*, *op. cit.*, pp. 108–109.

29. Cf. Dean's distinction: "The term *governmentality* seeks to distinguish the particular mentalities, arts and regimes of government and administration that have emerged since 'early modern' Europe, while the term *government* is used as a more general term for any calculated direction of human conduct" (Dean, *op. cit.*, p. 2).

any permanent meaning of the terms. Analytics of government directs our attention to

> thought as it is embedded within programmes for the direction and reform of conduct. The analysis of government is concerned with thought as it becomes linked to and is embedded in technical means for the shaping and reshaping of conduct and in practices and institutions. Thus to analyse mentalities of government is to analyse thought made practical and technical [...] The key starting point of an analysis of government is the identification and examination of specific situations in which the activity of governing comes to be called into question, the moments and the situations in which government becomes a problem.[30]

The Historical Development of Governmentality

Even though governmentality understood in its specific form has a continuous dynamic, three broad stages of the historical development of governmentality can be traced in Foucault and the commentaries to him, three different ways of couching governmental power, so to speak, relating to three broad "situations in which government becomes a problem".[31] The first is the one mentioned above, intertwined with the state's self-reflection and the beginning government of population, as developed from around 1580 and onwards towards 1650.[32] This is where Foucault sees the phenomenon of governmentality born from the Christian pastorate, the diplomatic-military technique and police (in the 17th-century sense of the term):[33] "With the sixteenth century we enter the age of forms of conducting, directing, and government."[34] Following Foucault, as quoted above, we could call this emerging polity "the administrative state",[35] and following Dean we will be stressing its "'dispositional' and 'householding' conception of government".[36] "Reason of state" writers developed theories that went above and beyond the traditional concern with the prince and his principality, and incorporated the ascending care for the self, the family and the state.[37]

30. *Ibid.*, pp. 18, 27.

31. From a Foucauldian perspective, periodisation is not uncontroversial. There are important continuities across the periods mentioned here, and changes were gradual, but even Foucault himself suggests that there were different forms of governmentality at different times, and so at least a rough periodisation would seem to be warranted. Since we assume domestic and inter-state governmentality to be interrelated, establishing some grip on the different periods matters to the ensuing analysis.

32. Foucault, *Security, Territory, Population, op. cit.*, p. 247.

33. *Ibid.*, p. 110. The discussion of these three phenomena takes up most of the 1978 lectures.

34. *Ibid.*, p. 231.

35. *Ibid.*, p. 109.

36. Dean, *op. cit.*, p. 6.

37. The seminal works on this remain Gerhard Oestreich, *Neostoicism and the Early Modern State* (Cambridge: Cambridge University Press, 1982); Richard Tuck, *Philosophy and Government 1572–1651* (Cambridge: Cambridge University Press, 1993). See also Dean, *op. cit.*, p. 85; Foucault, *Security, Territory, Population, op. cit.*, pp. 65, 88, 93. In later work, Tuck suggests that the direction should be seen as descending rather than ascending. According to him, the idea of the autonomous agent was first developed in thinking about international affairs, then brought "into *civil* life: all politics was now seen as at least potentially civil war, and our fellow citizens were no different from enemies with whom we lived in uneasy peace" (Richard Tuck, *The rights of war and peace: political*

In general it might be possible to speak of the development of a "state rationality",[38] with *reason of state*, police, polizeiwissenschaft, cameralism and mercantilism as different instantiations:

> "Œconomy" as a set of techniques of an art of government conceived as the wise administration of the state as royal household, police as a condition to be achieved equivalent to the good order of the city or territory, and population as the numbers of the people or the "stock of labour" that makes up the wealth, strength and greatness of the nation, are all the rudiments of a rationality of government that was also a rationality of the state in the seventeenth and eighteenth centuries.[39]

Or, to follow Foucault:

> The state is at once that which exists, but which does not yet exist enough. *Raison d'État* is precisely a practice, or rather the rationalization of a practice, which places itself between a state presented as given and a state presented as having to be constructed and built [...] To govern according to the principle of *raison d'État* is to arrange things so that the state becomes sturdy and permanent so that it becomes wealthy, and so that it becomes strong in the face of everything that may destroy it.[40]

Towards the end of the 18th century a new break can be identified, with the emergence of a liberal political economy and new thinking on population:

> A new governmentality is born with the *économistes* more than a century after the appearance of that other governmentality in the seventeenth century. The governmentality of the *politiques* gives us police, and the governmentality of the *économistes* introduces us, I think, to some of the fundamental lines of modern and contemporary governmentality.[41]

In Foucault's thinking, this governmentality is centred on society over and above population. In an immanent critique, Curtis points out that "population" in Foucault's accounts is a word that glosses over the differences between three different but linked concepts; populousness (when Foucault discusses police and mercantilism), the Social Body (when Foucault discusses biopolitics) and population (as Curtis thinks it should be applied, when Foucault discusses liberal government).[42] He then moves on to criticise how this conflation lends a teleological overtone to the analysis and places the development of liberal governmentality a century and a half too early. Finally, he points out that a genealogical account must take into consideration how population emerges as a gradual result of sovereign and disciplinary

thought and the international order from Grotius to Kant (Oxford: Oxford University Press, 1999), p. 10; emphasis in original). This is a reading that comes closer to Foucault's ideas in *"Society Must be Defended"*.

38. Dean, *op. cit.*, p. 92.
39. *Ibid.*, p. 95.
40. Foucault, *The Birth of Biopolitics*, *op. cit.*, p. 4.
41. Foucault, *Security, Territory, Population*, *op. cit.*, p. 348.
42. Bruce Curtis, "Foucault on Governmentality and Population: The Impossible Discovery", *Canadian Journal of Sociology*, Vol. 27, No. 4 (2002), p. 507.

applications of power. Dean argues somewhat along the same lines when he distinguishes between pre- and post-Malthusian conceptions of population,[43] and argues that post-Malthusian thinking on population coincides with the emergence of the gradual "governmentalization of the state", where earlier forms of sovereign and disciplinary power were transformed and supplemented by "the regulation of populations and individuals and the psychological, biological, sociological and economic processes that constitute them".[44] In short, what we see is the emergence of biopolitics and liberalism,[45] a "shift from a government of inhabitants, 'things' and households to a government through tendencies, laws, necessities and processes".[46]

And finally, we have over recent decades seen the governmentalisation processes reflexively turned back on the state, through a process that could be called "the governmentalisation of government". This is the form of governmentality that is most often implied in current studies, and it is thus relatively well known. Since the purpose of this article is to explore how a governmentality perspective can be applied historically to inter-state relations, we will not discuss the current form of governmentality any further here.

Inter-state Governmentality?[47]

Governmentality approaches over the last 15 years, starting with the publication of *The Foucault Effect*,[48] have largely dealt with the study of domestic societies. Nevertheless, a growing body of IR studies is applying a governmentality perspective.[49] Contributions of the last couple of years have even occasioned a review essay.[50]

While there are some examples of approaches tackling governmentality in broad terms,[51] most of the studies of global/international governmentality have been dealing with specific phenomena. Expanding the scope, Sending and

43. Dean, *op. cit.*, pp. 94–95 and Chapters 5 and 6.

44. *Ibid.*, p. 210. The publication of *Security, Territory, Population* and *The Birth of Biopolitics* makes it clear that Foucault was at least to some extent aware of the problems raised by Curtis and Dean.

45. Foucault spends the first three lectures of the 1979 series on this emergence. Dupont and Pearce suggest that in Foucault's account this liberal governmentality becomes a sort of telos, which must overcome blockages to realise itself; Danica Dupont and Frank Pearce, "Foucault contra Foucault: Rereading the 'Governmentality' Papers", *Theoretical Criminology*, Vol. 5, No. 2 (2001), pp. 123–158. This critique of an "objective idealism" is a lot less potent after the publication of Foucault's entire lecture series. Their critique of "subjective idealism", the reliance on formal discourses in a few central countries and an intentional reading of them is weakened, but still has some merit. It should nevertheless be noted that the lectures were a first stab, an opening up of a field for research.

46. Dean, *op. cit.*, p. 111.

47. By now it is perhaps obvious, but the choice of "inter-state" is intended to avoid both the anachronism (when dealing with early modern Europe) of "international" and the unwarranted (at the time) "global".

48. Graham Burchell, Colin Gordon and Peter Miller (eds.), *The Foucault Effect: Studies in Governmentality* (Chicago, IL: University of Chicago Press, 1991).

49. A useful, if already somewhat dated, overview can be found in Wendy Larner and William Walter, "Introduction: Global Governmentality", in Wendy Larner and William Walter (eds.), *Global Governmentality: Governing International Spaces* (Milton Park: Routledge, 2004).

50. Michael Merlingen, "Foucault and World Politics: Promises and Challenges of Extending Governmentality Theory to the European and Beyond", *Millennium*, Vol. 35, No. 1 (2006), pp. 181–196.

51. For example, Mitchell Dean, "*Nomos* and the Politics of World Order", in Larner and Walter (eds.), *op. cit.*; Barry Hindess, "Liberalism—What's in a Name?", in Larner and Walter (eds.), *op. cit.*

Neumann eloquently defend the application of governmentality to the study of the international realm in general.[52] They share with most other writers a predilection for the (neo)liberal form(s) of governmentality, particularly the governmentalisation of government. To the extent that they make the usefulness of the perspective dependent on the steady thickening of "the international" through liberal norms and increasing numbers of non-state actors, they seem to be unduly restricting its application, in time and in scope.[53] However, in their forthcoming work, Neumann and Sending move further back in time than they have previously done, and look into how governmentality perspectives can shed light on what constitutes a great power.[54] In this, they are inspired by Barry Hindess's comments on how the liberal government of domestic populations presupposes a world divided into territorial states, carving up the global population into manageable entities.[55] Although Hindess does refer to the Peace of Westphalia, it is nonetheless clear that he refers mainly to liberal governmentality, more specifically the post-1815 period. Devetak pushed the timeframe even further, making a good argument about how Foucauldian perspectives could enrich our understanding of early modern state building, and enhance the accounts of historical sociology.[56] This seems obviously true,[57] but is still confined to the insides of states, not the relations between them. In general, very little work has yet been undertaken on the period before the Second World War, and scarcely any on the period from *c.*1600 to *c.*1800.

Foucault on Inter-state Governmentality

A preference for (neo)liberal governmentality in the literature is understandable, first since most scholars engage with current themes, and second because the

52. Ole Jacob Sending and Iver B. Neumann, "Governance to Governmentality: Analyzing NGOs, States, and Power", *International Studies Quarterly*, Vol. 50, No. 3 (2006), pp. 651–672; Iver B. Neumann and Ole Jacob Sending, "'The International' as Governmentality", *Millennium*, Vol. 35, No. 3 (2007), pp. 677–701.

53. Such a bias is even more obvious in the general literature; cf. the books and themes covered by Merlingen, *op. cit.*

54. Iver B. Neumann and Ole Jacob Sending, *Governing the Global Polity: World Politics as Governmentality* (Ann Arbor: University of Michigan Press, forthcoming 2010).

55. See for example, Hindess, "Liberalism", *op. cit.*; *idem*, "Government and Discipline", *International Political Sociology*, Vol. 2, No. 3 (2008), pp. 268–270.

56. Richard Devetak, "Foucault, Discipline and *Raison d'État* in Early Modern Europe", *International Political Sociology*, Vol. 2, No. 3 (2008), pp. 270–272.

57. Particularly if we compare Foucault to the predominantly materialist accounts of much historical sociology, e.g. Charles Tilly, "Reflections on the History of European State-making", in Charles Tilly (ed.), *The Formation of National States in Western Europe* (Princeton: Princeton University Press, 1975); *idem*, *Coercion, Capital, and European States, AD 990–1992* (Cambridge, MA: Blackwell, 1995); Michael Mann, *The Sources of Social Power*, Vol. I (Cambridge: Cambridge University Press, 1986). Writing at roughly the same time as Foucault, Oestreich lamented that "The elaboration of army organization and state finance, two of the most important instruments at the state's disposal in the sixteenth and seventeenth centuries, are held to have resulted from military and political necessity and to have evolved by themselves in response to the requirements of the real world [...] Hence, the theories of practical government which were current at the time have been left largely unexamined, as opposed to those which are interesting from the standpoint of legal and constitutional philosophy" (Oestreich, *op. cit.*, p. 36). Later historical sociology has incorporated a larger focus on cultural factors, but a Foucauldian perspective, with an analysis of discourse that incorporates the study of practice and materiality would still seem to have a lot to add to this literature.

previously published works by Foucault paid far more attention to (neo)liberal governmentality than to its predecessors. The publication of *Security, Territory, Population* and *The Birth of Biopolitics*, on the other hand, provides us with a much wider and more historically grounded approach, leading Valverde to suggest that Foucault is pointing in the direction of "Genealogies of European states".[58]

When reading the material surrounding the already published lecture in which governmentality was introduced, one quickly sees that the international is crucial to Foucault's analysis, in ways that should resonate with IR scholars. It was noted above that "the new diplomatic-military technique" was seen by Foucault as one of the cornerstones of the governmentalisation of the state. The theme arises in the larger context of *reason of state*. For Foucault the inroad is a reading of Botero, which leads him to comment on "a very important phenomenon. It is that states are situated alongside other states in a space of competition. And I think that at the time this idea was at once fundamental, new, and extremely fruitful with regard to everything that we may call political technology."[59] Rivalries, wars and struggle had obviously existed previously, but again the critical point comes with self-reflexivity—the ability to conceive of one's own state as one of many. With the move from princely rivalries to state competition, force becomes absolutely central, and thus also maintaining an equilibrium or balance between forces.[60] To handle this, the states set up "assemblages"; a military-diplomatic apparatus to take care of relations of force between the entities and the apparatuses of police broadly understood to enhance force within the entities.[61]

Starting with the military-diplomatic apparatuses, the objective is the balance of Europe, allowing states to maximise their forces without upsetting the system as such or bringing on its own ruin. To the extent that there is a goal to this balancing, it is universal peace. But where universal peace had used to be seen as the result of universal empire (secular or religious), it would now come from "plurality maintained as plurality".[62] The instruments for maintaining this balance were first war, divested from juridical consideration and steeped in politics. The second

58. Mariana Valverde, "Genealogies of European States: Foucauldian Reflections", *Economy and Society*, Vol. 35, No. 1 (2007), pp. 159–178. Valverde presents a rich and insightful reading of *"Society Must be Defended"* and *Security, Territory, Population*, but does not reflect on the development in Foucault's thought from war/battle to government, and, like Devetak, does not really engage with the inter-state level.

59. Foucault, *Security, Territory, Population*, *op. cit.*, p. 289.

60. It is notable that post-structural accounts of foreign policy, like that presented by David Campbell, had developed a quite similar model for the co-constitutive emergence of states and state system through foreign policy acting as a wall (rather than a bridge); David Campbell, *Writing Security: United States Foreign Policy and Politics of Identity*, 2nd edn (Minneapolis: University of Minnesota Press, 1998). However, as noted by Felix Berenskoetter, such views, and indeed most of IR scholarship and modern liberal thought, rest on an ontologically unfounded assumption about necessary enmity between people and units, going back at least to Hobbes. Drawing on Heidegger and Aristotle, Berenskoetter suggests a completely different narrative, where states seek recognition and stabilisation of Self, and where friendship rather than enmity is the normal relation; Felix Berenskoetter, "Friends, There Are No Friends? An Intimate Reframing of the International", *Millennium*, Vol. 35, No. 3 (2007), pp. 647–676. In such a perspective, "friendship" (or, for that matter, enmity) is analytically neither prior nor subsequent to the establishment of contact between units, they develop analogously. Foucault quite clearly falls into the "enmity" category, by stressing how states by necessity react against the increasing force of other states by increasing their own force.

61. Foucault, *Security, Territory, Population*, *op. cit.*, pp. 295–297.

62. *Ibid.*, pp. 297–300.

instrument was permanent diplomacy of a self-reflecting kind: "there is the idea of a permanent apparatus of relations between states that is not an apparatus of imperial unity or ecclesiastical universality. It is the idea of a veritable society of nations, and I am not employing a word retrospectively here."[63] With a nod to Crucé, Foucault notes that there circulates an "idea that between themselves states form something like a society in the European space, the idea that states are like individuals who must have certain relations between them", and that this idea was what gave rise to the law of nations.[64] The third and final instrument was the deployment of a permanent military apparatus.[65] To put it schematically, at the general level we have reason of state.[66] This can be subdivided into, on the one hand, the military-diplomatic apparatuses concerned with maintaining the balance of power (war, diplomacy/society of nations/the law of nations and the permanent military apparatus), on the other, "police" in the 17th-century meaning of the term.

Reason of state. We touched upon the internal aspects of reason of state above. Moving on to the external aspects, it should be noted straight away that in the passages dealing with reason of state and the relationship between the state and the other states, Foucault jumps freely back and forth in time, and the empirical references are sparse. As Sennellart notes,[67] Foucault's knowledge of the secondary literature on raison d'état was somewhat limited. The quotations are furthermore scattered from the early 17th to the late 18th century. If we are interested in exploring what an application of a governmentality perspective to inter-state relations would look like, and we assume that the domestic form of governmentality is interrelated to the inter-state form of governmentality, it matters, at least analytically, whether the phenomena under discussion should be understood in connection with the system of administrative states or with the gradual governmentalisation of the states.

Furthermore, the mixing and matching that Foucault engages in raises an absolutely central and critical point regarding the conception of time during the period

63. *Ibid.*, p. 303.

64. *Ibid.*

65. One could be forgiven for speculating what would have happened if Foucault had read Headley Bull, as these themes are very similar to the central institutions described by English School authors; Hedley Bull, *The Anarchical Society* (Houndmills: Macmillan, 1977). It could easily be argued that one of the main contributions of the "first wave" of post-structuralist IR was to juxtapose Foucauldian (and other post-structural) thought with the alleged canon of IR (including Bull), so as to deconstruct, destabilise, denaturalise and historicise knowledge of the international. See, for example, Richard K. Ashley, "The Geopolitics of Geopolitical Space: Towards a Critical Social Theory of International Politics", *Alternatives*, 12 (1987); James Der Derian, *On Diplomacy: A Genealogy of Western Estrangement* (Oxford: Blackwell, 1987); R.B.J. Walker, *Inside/Outside: International Relations as Political Theory* (Cambridge: Cambridge University Press, 1993); Timothy W. Luke, "Discourses of Disintegration, Texts of Transformation: Re-reading Realism in the New World Order", *Alternatives*, 18 (1993); *idem*, "Governmentality and Contragovernmentality: Rethinking Sovereignty and Territoriality after the Cold War", *Political Geography*, Vol. 15, Nos. 6/7 (1996); Gearoid Ó Tuathail, *Critical Geopolitics: The Politics of Writing Global Space* (Minneapolis: University of Minnesota Press, 1996). While presenting forceful interventions against the current (at the time) practice of (Realist) IR, only some of these authors engaged historical material directly, and only some of them applied a governmentality perspective. To my knowledge, there have been no explicitly governmentality-inspired analyses of any part of the early modern European state system.

66. We will leave pastoral power out of the discussion.

67. Sennelart, *op. cit.*, p. 398, n. 98.

of the administrative state, and the historical possibility of imagining many of the phenomena that Foucault finds during the period. Put briefly, he seems to have underestimated the continued importance of a cyclical and religious view of history, and the extent to which a system of equal states was even conceivable in the early modern period.

The point can be illustrated by looking at the purported secularity and newness of the early modern (administrative) state. To Foucault during this period, "The state exists only for itself and in relation to itself, whatever obedience it may owe to other systems like nature or God. The state only exists through and for itself."[68] Even though antecedents can be found, the early modern state and system of states was codified with the Peace of Westphalia in 1648, and, if we are to believe Foucault, these states were fundamentally different from the preceding ones:

> in its foreign policy, let's say in its relations with other states, the state, or rather government according to *raison d'État*, has a limited objective in comparison with the ultimate horizon, the project and desire of most sovereigns and governments in the Middle Ages to occupy the imperial position with regard to other states so that one will have a decisive role both in history and in the theophany. *Raison d'État*, on the other hand, accepts that every state has its interests and consequently has to defend these interests, and to defend them absolutely, but the state's objective must not be that of returning to the unifying position of a total and global empire at the end of time. It must not dream that that one day it will be the empire of the last day. Each state must limit its objectives, ensure its independence, and ensure that its forces are such that it will never be in an inferior position with respect to the set of other countries, or to its neighbors, or to the strongest of all the other countries [. . .] military-diplomatic policy is organized by reference to the principle of the state's self-limitation, to the principle of the necessary and sufficient competition between different states.[69]

There are several comments to be made here. The argument that universal empire was not a goal would seem to contradict the most common reading of the foreign policies of France during most of this period, and to underestimate the extent to which any balancing in the system was related to counteracting just such aspirations to dominance. The mutual recognition of states' interests did indeed matter in inter-state relations, in that it fostered predictability and reduced more "passionate" behaviour,[70] but if we follow Hirschman, the adoption of "interests" were gradual and much more relevant towards the end of the era of the administrative

68. Foucault, *The Birth of Biopolitics*, *op. cit.*, p. 5. According to Foucault, *raison d'État* "was broadly formed during the sixteenth century" (*ibid.*, p. 4), but the assertion that the state was self-sufficient stands quite clearly at odds with the thought of one of the central figures of late 16th/early 17th-century reason of state, Justus Lipsius. Justus Lipsius, *Sixe Bookes of Politickes or Civil Doctrine*, translated into English by William Jones (Amsterdam: Theatrum Orbis Terrarum, [1594]1970). Cf. Oestreich, *op. cit.*; Halvard Leira, "Justus Lipsius, Neostoicism and the Disciplining of 17th Century Statecraft", *Review of International Studies*, Vol. 34, No. 4 (2008), pp. 669–692.

69. Foucault, *The Birth of Biopolitics*, *op. cit.*, p. 6.

70. A.O. Hirschmann, *The Passions and the Interests: Political Arguments for Capitalism before its Triumph* (Princeton: Princeton University Press, 1977), p. 51.

state than at its beginning. Of even higher importance is how Foucault exaggerates the extent to which God is taken out of the equation.

If we turn instead to the German conceptual historian Reinhart Koselleck, a picture emerges in which the fundamental conceptual shifts in the Western political language took place roughly between 1750 and 1850 (the so-called "saddle-time"). This was not only a period that saw fundamental conceptual change but also the period where history itself became, so to speak, historicised:

> Formerly there had existed, for instance, the history that God had set in motion with humanity. But there was not history for which humanity might have been the subject or which could be thought of as its own subject. Previously, histories had existed in the plural—all sorts of histories had occurred and might be used as exempla in teachings on ethics and religion, and in law and philosophy [. . .] this plural form was modified into an objectless singular. One of the conceptual achievements of the philosophy of the enlightenment was enhancing history into a general concept which became the condition of possible experience and possible expectation. Only from around 1780 can one talk of "history in general".[71]

Only then was it possible for history, as we now think of it, to be made. This assessment stands in stark contrast with Foucault's claim about the earlier period that "In fact, the plurality of states is the very necessity of a history that is now completely open and not temporally oriented towards a final unity. The theory of *raison d'État* I talked about last week entails an open time and a multiple spatiality."[72] On the contrary, time was not yet open. Koselleck also goes into more detail regarding the novelty of reason of state thought. He argues that ideas about the future had indeed moved from prophecy to prognosis, but adds that the future was still bound by dynasticism and the human lifespan:

> Based as it was on the life and character of acting personages, the European republic of rulers could still understand history in natural terms. It is not surprising that the ancient pattern of cycles put back in circulation by Machiavelli found such general support. This experience of history, founded as it was on repeatability, bound prospective futures to the past. This certainly makes clear that the distance separating the early modern political consciousness of time from that of Christian eschatology was nowhere as great as it might seem. *Sub specie aeternitatis* nothing novel can emerge, whether the future is viewed in terms of faith, or of sober calculation. A politician could become more clever or even cunning; he could refine his technique; he could become wiser or more farsighted: but history never conveyed him into unknown regions of the future. The reoccupation of a prophesied future by a predicted future had not yet fundamentally ruptured the plane of Christian expectations. That is what harnesses the republic of rulers to the Middle Ages, even if it no longer conceives of itself as Christian.[73]

71. Reinhart Koselleck, *Futures Past: On the Semantics of Historical Time* (New York: Columbia University Press, 1985), p. 194.
72. Foucault, *Security, Territory, Population, op. cit.*, p. 290.
73. Koselleck, *op. cit.*, p. 21, cf. p. 197.

Even though final unity in empire might have receded as the ultimate goal, God still had to be taken into account, and, more importantly, history itself remained cyclical.

Foucault also seems to overestimate the degree to which reason of state thought dealt systematically with external affairs. Even though issues like empire, war and international law are treated, reason of state is, as Foucault himself had noted, first and foremost concerned with the "business of rule",[74] the governing of self, family and state. It concerns itself first and foremost with the inside of states and with how they project outwards, not with foreign policy per se as we would think of it today.[75] In the period of the administrative state, the examples of system-level thought are scarce, often related to empire, and brief. From the middle of the 18th century, Enlightenment philosophers like Rousseau, Paine and Kant did engage somewhat more methodically with system-level thought, but by that time we are well into the period of the governmentalising states. To an even larger degree than Foucault emphasised, the order that emerged between states was the result not of systems-level thinking but the doings of individual states.[76] This does not imply giving states primacy over the state system, but indicates that the ensuing system was not a system of like, sovereign units.

"The state only exists as states, in the plural",[77] Foucault claims in a typically concise way, while capturing the idea of the co-constitution of the state and the state system, in the context of the historical relationship between the emergence of this co-constitutive relationship and the gradual decline of universal empire. However, this plurality was a plurality of different entities. In several publications, Andreas Osiander has argued, on the one hand, that 1648 was a lot less important than we are wont to think, and on the other that what characterised the post-Westphalian system was not so much sovereignty as autonomy.[78] Bartelson argues along the same lines when he states that an "international system" did not exist as an object of knowledge at the time, but that what could be found was "a *tabulated order of states*" (emphasis in original).[79] Foucault's argument that "the idea that states are in competition with each other is basically the direct, almost ineluctable consequence of the theoretical principles posited by *raison d'État*"[80] in this respect seems like yet another reiteration of what has been dubbed the "Westphalian Myth" in IR; the reading backwards of 19th-century power politics into the 17th and 18th centuries, and the idea that something recognisable as the

74. I owe this formulation to Benjamin de Carvalho.

75. Cf. Jens Bartelson, *A Genealogy of Sovereignty* (Cambridge: Cambridge University Press, 1995). It is worth noting that "foreign affairs" can be found in use in English as early as 1611, but that "foreign policy" would not appear until 1804, perhaps suggesting that whereas the polity could have interactions with like units in the early 17th century, regularised and planned interaction would have to wait until the early 19th century.

76. An obvious caveat must be mentioned immediately; the European aristocracy (to some extent including royalty) was to a large degree cosmopolitan, and were almost in sole control of diplomacy and higher military positions.

77. Foucault, *The Birth of Biopolitics*, *op. cit.*, p. 5.

78. See, in particular, Andreas Osiander, *The States System of Europe, 1640–1990* (Oxford: Oxford University Press, 1994).

79. Bartelson, *op. cit.*, pp. 137–139.

80. Foucault, *Security, Territory, Population*, *op. cit.*, p. 290.

modern state system came into being in 1648.[81] Competition might follow from the domestic policies of states, but systems-level thinking was simply not very well developed during this period.[82] Three implications arise from this. First, inter-state governmentality needs to be periodised in a way similar to the governmentality of states. Second, exploration of inter-state governmentality through reason of state requires a reading attuned to the continued importance of a religious and cyclical view of history. Third, an illumination of changing inter-state governmentality cannot rely solely on a static reading of reason of state, but must be dynamic and incorporate other literatures and practices as well.

Balance of power. The previous discussion obviously also has a bearing on Foucault's take on the balance of power. As noted above, to Foucault, balancing emerges as a result of the gradual move from dynastic rivalries to stately competition, a move that was closely connected to the growth of force as a central concept. Foucault reads the move from rivalry to competition as finished by the Treaty of Utrecht (1713), but believes it to be traceable with practitioners already in the first half of the 17th century. Unless we insist on prioritising enmity over friendship, this assertion stands somewhat at odds with the observation that the idea of states being "friends", or partaking in "friendly" relations, stems from around 1800; until then, the friends were individual kings and/or lords.[83]

To Foucault, the balance of power is the key mechanism maintaining not only the external balance but also the combination of limited and limitless goals:

> Competition between states is precisely the hinge connecting these limited and unlimited objectives, because it is precisely so as to be able to enter into competition with other states, that is to say, maintain an always uneven, competitive equilibrium with other states, that government [has to regulate the life of] its subjects, to regulate their economic activity, their production, the price [at which] they sell goods and the price at which they buy them, and so on [...] The correlative of this limitation of the international objective of government according to *raison d'État*, of this limitation in international relations, is the absence of a limit in the exercise of government in the police state.[84]

To some extent, Foucault seems to be contradicting himself. We saw above that he claimed that "the idea that states are in competition with each other is basically the direct, almost ineluctable consequence of the theoretical principles posited by

81. On this see, in particular, the works of Krasner and Osiander: Stephen Krasner, "Westphalia and All That", in Judith Goldstein and Robert Keohane (eds.), *Ideas and Foreign Policy* (Ithaca, NY: Cornell University Press, 1993); *idem*, "Compromising Westphalia", *International Security*, Vol. 20, No. 3 (1995/96), pp. 115–151; Osiander, *op. cit.*; *idem*, "Sovereignty, International Relations and the Westphalian Myth", *International Organization*, Vol. 55, No. 2 (2001), pp. 251–287.

82. It should be noted that Foucault does acknowledge that some states are "more equal" than others, but the regulating framework is still said to be sovereignty; Foucault, *Security, Territory, Population*, *op. cit.*, pp. 297–298.

83. Evgeny Roshchin, "The Concept of Friendship: From Princes to States", *European Journal of International Relations*, Vol. 12, No. 4 (2006), p. 615.

84. Foucault, *The Birth of Biopolitics*, *op. cit.*, p. 7.

raison d'État",[85] i.e. that states in a state of competition followed from reason of state logic. In the longer quote, the competition and the ensuing equilibrium seems to be what promotes the police state at home. As indicated above, any co-constitution of state and order of states at this time had to be based on ideas about hierarchy and autonomy, rather than anarchy and sovereignty. Furthermore, if we accept the idea of the administrative state as largely householding, incorporating self, family and state in an ascending relationship, external balancing could hardly be defining the internal workings of the state. If, by way of example, we turn to one of the seminal early writers on the issues at hand, Justus Lipsius,[86] who informs Oestreich's very influential account of the conceptual and ideological underpinnings of the emergence of the early modern state,[87] the development of the police state is largely disconnected from external affairs, and concerned with creating internal order (e.g. to avoid civil strife).

What, then, about balancing as such? Foucault believes it to have been finally instituted in 1648, implying by the term:

> The absolute limitation of the force of the strongest, the equalization of the strongest, and the possibility of the combination of the weaker against the stronger are the three forms conceived and devised to constitute European equilibrium, the balance of Europe.[88]

Foucault is certainly correct in pointing out the use of balancing terms in the literature of the time, but his tripartite division does not capture some of the other critical questions regarding balancing. We could, for example, ask whether the balancing was intuitive (an unreflective practice), subjective or intersubjective, whether it was conceived as a law-like regularity or as an empirical counterfactual, whether it was conceived as bipolar or multipolar and whether it was conceived in adversarial or associational terms.[89] Current scholarship would tend to date self-reflective balancing practices based on multipolarity, and not only simple balancing, later than Foucault, around the Peace of Utrecht rather the Treaties of Westphalia.[90] Foucault is clearly overstating the case when discussing "the laws of equilibrium" as active in balancing practices in 1648. However, the distinct associative traits that Foucault describes in the balancing practices accord well with current readings,[91] which de-emphasise the adversarial traits as an anachronistic overlay from the 20th century.

As for the instruments of the balance, Foucault has relatively little to add on warfare,[92] which he sees as increasingly divested from juridical consideration, and steeped in politics. The claim is largely unsubstantiated and seems odd, to say the least, when considering the continued evolvement of just war-thought

85. Foucault, *Security, Territory, Population, op. cit.*, p. 290.

86. Lipsius, *op. cit.*

87. Oestreich, *op. cit.*

88. Foucault, *Security, Territory, Population, op. cit.*, p. 299.

89. Richard Little, *The Balance of Power in International Relations: Metaphors, Myths and Models* (Cambridge: Cambridge University Press, 2007); Torbjørn L. Knutsen, "The Rise of Balance-of-power as an Ordering Institution", Paper presented at the 48th ISA convention, Chicago, Illinois, 2007.

90. Osiander, *The States System of Europe, op. cit.*; Knutsen, *op. cit.*

91. Little, *op. cit.*

92. Foucault, *Security, Territory, Population, op. cit.*, pp. 300–302.

in this period, when justness was increasingly divested from divine law and rooted in natural law.[93]

There is a little more substance to Foucault's discussion of diplomacy,[94] which included the idea of a society of states and international law, even though permanent diplomacy obviously pre-dated the Treaties of Westphalia considerably, contrary to what Foucault implies, as did the law of nations. There is the added problem of dating the idea of a society of states. Foucault argues that it can be found in Crucé, but, as the notes make clear,[95] this simply seems to be a misunderstanding on his part. Foucauldian-inspired scholars of international law would dispute that the idea that the states as a system of "like individuals" (as quoted above) gave rise to the law of nations, stressing instead the genealogical continuities and how the confrontation with the radical (Amerindian) Other might have been just as important as the self-reflexivity of like units. In this perspective, difference in the form of radical Otherness was more important than difference between similar entities.[96]

Turning finally to the permanent military apparatus, Foucault is largely in line with current research,[97] even though it should be added that it had both its practical and its intellectual roots around 1600, and was part and parcel of the development of the governmentality of the administrative state.[98]

Before leaving the primary sources, a few comments have to be made about police. In the discussion about police there is necessarily less focus on the outside of the state.[99] Nevertheless, there are connections. Most noteworthy are the connections between equilibrium and police, as alluded to above:

> One can only effectively maintain the balance and equilibrium in Europe insofar as each state has a good police that allows it to develop its own forces [. . .] In the end there will be imbalance if within the European equilibrium there is a state, not my state, with bad police. Consequently, one must see to it that there is good police, even in other states. European equilibrium begins to function as a sort of inter-state police or as right. European equilibrium gives the set of states the right to see to it that there is good police in every state. This is the conclusion drawn explicitly and systematically in 1815 with the Vienna treaty and the policy of the Holy Alliance.[100]

If states cannot police their population, someone else will have to do the job. The date 1815 is particularly noteworthy, as by then we are well into the liberal period

93. See, for example, Anthony Anghie, *Imperialism, Sovereignty and the Making of International Law* (Cambridge: Cambridge University Press, 2005); Halvard Leira, "At the Crossroads—Justus Lipsius and the Early-modern Development of International Law", *Leiden Journal of International Law*, Vol. 20, No. 1 (2007), pp. 65–88.

94. Foucault, *Security, Territory, Population, op. cit.*, pp. 302–305.

95. *Ibid.*, pp. 309–310, nn. 26–30.

96. Anghie, *op. cit.*

97. Foucault, *Security, Territory, Population, op. cit.*, pp. 305–306.

98. Oestreich, *op. cit.*; Leira, "Justus Lipsius, Neostoicism and the Disciplining of 17th Century Statecraft", *op. cit.* and works quoted therein.

99. On "police", see in general Oestreich *op. cit.*

100. Foucault, *Security, Territory, Population, op. cit.*, pp. 314–315. Foucault gives the Concert of Europe as an example of this, but one is tempted also to relate it to "a balance of power that favors freedom".

of governmentalisation of the state. Thus, the external aspect of police, one of the key building blocks of the administrative state, is seen to have come to full fruition only after a new type of governmentality has emerged. There is no inherent contradiction in one technique of power surviving under a new governmentality, quite the contrary, but the argument that external "policing" was critical to the balancing of the administrative states cannot be made retroactively based on the situation after 1815.

Conclusion—Taking Foucault beyond Foucault

Although the publication of Foucault's lecture series shows a greater interest in inter-state relations than what can be found in his regular publications, his comments are sparsely referenced, conceptually ambiguous and dangerously eclectic when moving from specific authors at rather different points in time to general tendencies. In particular, Foucault fails to differentiate properly between what should logically be seen as elements of liberal inter-state governmentality and the inter-state governmentality of the administrative states. This de-differentiation is particularly problematic since Foucault does not really reflect on the massive conceptual changes that took place around the end of the 18th century, including the emergence of the historicity of history, changes which enabled ideas about progress and thus system-wide change.[101] These problems in Foucault's account necessitate both a careful approach when applying Foucauldian insights and further clarification of what and when can be studied.

One potential objection can be discarded straight away. Based on Foucault's comments on (neo)liberal governmentality, an implicit argument in much governmentality research at the global level seems to have been that such research is dependent on the gradual emergence of a global society, to mirror the earlier emergence of domestic societies. With the elaboration of the earlier governmentality of the administrative state, such objections can be discarded. We should in any case be careful to avoid undue domestic analogies, and government/governmentality, whether approached as a type of power relation, a technique or a mechanism, is not logically dependent on the field of application. It does, however, seem to depend on the self-reflectivity of a polity, and thus drawing the line around 1600, as Foucault does, makes sense.[102]

However, it could be argued that the self-reflectivity of the state was not followed by self-reflectivity about being part of a system. Whereas government in a sense became self-conscious around 1600, and political science in its earliest forms was established at this point, systematic knowledge about an "international" system would only emerge from around 1800, and IR as a discipline obviously even a century later. The answer to such an objection could follow Bartelson in stressing that there was indeed a consciousness about states being

101. The de-differentiation suggests that Foucault might at some level have built on the distinction between inside and outside that has been deconstructed in IR over the last two decades (see, in particular, Walker, *op. cit.*), particularly since he differentiates much more clearly between the periods of intra-state governmentality, and since the changes between the epistemes of the different ages in question here are central in his earlier writings (such as *The Order of Things*).

102. Although not a categorical one—we should expect to find self-reflective governmental practices earlier as well, although not in systematic patterns.

part of "a *tabulated order*",[103] and that even though there was little systemic reflection, the problem of how to interact with relatively similar polities did arise in a systematic way for the first time in this period. Thus, and this is the first conclusion, it does indeed make sense to apply governmentality perspectives to inter-state relations even in the period of the administrative state (*c*.1580–1800).

Moving on to what one should be looking for, the need again arises to push the Foucauldian approach past Foucault. It is striking how, to Foucault, power in the inter-state system is not productive to the same extent as domestically. As he comments, the diplomatic-military apparatus "has hardly changed since the eighteenth century".[104] This is a claim that that sits uneasily not only with much theorising about international affairs but also with other Foucauldian observations about seeming stability. Remember the methodological self-description:

> I wanted to reintegrate a lot of obvious facts of our practices in the historicity of some of these practices and thereby rob them of their evidentiary status, in order to give them back the mobility that they had and that they should always have.[105]

The very seeming stability of the instruments of the diplomatic-military apparatus should lead us to question them, if only to discover how they have remained stable. It is telling that one of the first IR texts to fully embrace a genealogical account recounts a story of diplomacy that is anything but stable over any length of time.[106]

Thus, investigation could fruitfully be directed at the very issues that Foucault mentioned: reason of state, balance of power, war, diplomacy and the standing military apparatus. The ways to go about this are manifold, and include applying Foucauldian approaches more systematically to these matters than Foucault himself did. In his genealogy of sovereignty, Bartelson largely followed this route, and proved to be a more thorough Foucauldian than Foucault himself when it comes to inter-state relations. Writing genealogies of key terms, such as balance of power or foreign policy, could thus be one way of approaching the subject matter. Such studies could fruitfully be combined or contrasted with approaches that draw on the more agency-centred *Begriffsgeschichte* tradition of conceptual history associated with Koselleck, or the Cambridge School of Skinner and Pocock.[107] Both the German and the British Schools of conceptual history already provide treasure-troves of empirical data for those seeking to think through early modern European inter-state relations from a governmentality perspective.

What, then, to look for? If one follows Foucault, two starting points come immediately to mind. The first is the one suggested several times already—to explore how certain problems arose, in the way that they did, at the time that they did. To name just a few examples: how did the notion of a comprehensive, subjective and objective balance of power gain ground; how did it become

103. Bartelson, *op. cit.*, pp. 137–139.

104. Foucault, *op. cit.*, *Security, Territory, Population*, p. 354.

105. Foucault, *op. cit.*, "What Our Present Is", pp. 138–139.

106. Der Derian, *op. cit.*

107. A good introduction can be found in Melvin Richter, *The History of Political and Social Concepts: A Critical Introduction* (Oxford: Oxford University Press, 1995).

necessary to codify diplomacy in manuals and how did the revitalisation of Roman military organisation come about? The second starting point gets at the same types of issues by asking how individuals governed and were governed by their own verities; how, for example, practices of diplomacy and warfare became taken for granted. Based on the above assertion that Foucault lacks a comprehensive grasp of what made the inter-state order of the 17th and 18th centuries hang together, closer attention could be paid in this perspective to manuals of diplomacy and warfare. The textbooks for diplomats are a good example of the "conduct of conduct"; detailing, for example, what sort of dissimulation was acceptable in public life, while still operating within the boundaries of Christian morality. The manuals of warfare combine the indirect conduct of conduct with more explicitly disciplinary themes. Diplomats and higher officers also had the strong common bond, across states, of nobility. As Foucault notes, raison d'état as well as the apparatus of equilibrium were "common notions or apparatuses found in most European countries, obviously with modulations".[108] This point is strikingly parallel to that furthered by Richard Ashley regarding the geopolitics of geopolitical space and the international community consisting of realist statesmen and theorists.[109] It also suggests the importance of the spread and uptake of notions and apparatuses, and how they became common. Studies of noble connections and ideas of what constituted proper noble conduct would thus also add to the picture. Finally, work could build further on Bartelson, and study "how policy became foreign", how the state governing itself also led to a governed order and how the possibility of foreign policy gradually emerged and made possible the reconceptualisation of the inter-state order as a system. While inspired by Foucauldian insights, concepts and methods, studies like these would go well beyond Foucault himself, and help elucidate not only the phenomena themselves but also the larger genealogies of the European states and the European state system.

108. Foucault, *Security, Territory, Population*, *op. cit.*, p. 316.
109. Ashley, *op. cit.*

Decentring Global Power: The Merits of a Foucauldian Approach to International Relations

DOERTHE ROSENOW

In recent times, the value of a critical approach to the study of International Relations (IR) that makes use of the concepts and methods of Michel Foucault has (again) been put on trial. I will argue in this article that both Foucauldians and their critics often neglect Foucault's radical epistemology that always prioritises practices over political theory. The demand of such an approach is the relentless decentring and diversifying of totalising and unifying accounts of (global) power relations, resulting in a continuous challenge of the traditional meta-theories and concepts of any academic discipline—including IR. The present article will follow this approach and challenge, through the investigation of a particular case of what is commonly perceived as an exercise in "global governance", the idea that contemporary (global) power relations can be depicted solely through the lens of neoliberalism, sovereignty, or biopolitics. Instead, it will show that (global) power is located in a complex and flexible constellation of diverse and contradictory, mutually constituting and mutually destabilising strategies and tactics at particular sites.

In recent times, the value of a critical approach to the study of International Relations (IR) that makes use of the concepts and methods of Michel Foucault has (again) been put on trial. I will argue in this contribution that both Foucauldians and their critics in IR too often neglect Foucault's radical approach to political power that constantly aims to decentre and diversify what seems uniform, as well as his particular epistemological standpoint that always prioritises practices over general theory. This approach continuously challenges traditional concepts of academic disciplines—including IR.

As I will show, theorists who criticise this take on IR from the critical/Marxist side often (involuntarily) close ranks with Foucauldians in their tendency to solidify and take for granted certain paradigms of global political rule, which leads to the problematic impression that "the international" is determined by a single overarching project. In contrast, this article will attempt to follow the Foucauldian path. Based on the investigation of a specific set of political practices, it will challenge three paradigms that are used particularly in critical and Foucauldian IR theory to decipher international power, namely neoliberalism, sovereignty, and biopolitics.

My engagement with the contemporary critique of a Foucauldian approach to IR in the first section of this article will be based primarily on David Chandler's recent contribution in *International Political Sociology*.[1] I will show how an in-depth understanding of Foucault's genealogical project reveals specific problems in Chandler's approach to questions of politics, power, and rights-based conceptions of political community. In the second section I will then attempt to clarify certain Foucauldian concepts to advance the central idea of the heterogeneity of power. I will point out the problematic connotations that are implied in Foucault's use of the metaphor of the governmental "triangle", and his concept of the "strategy", and will argue that Foucault's notion of the *dispositif* is better equipped to serve his project of decentring.

The third, fourth and fifth sections will then engage with a particular set of political practices to be found at the site of the World Trade Organization (WTO). The case I will examine is the EC Biotech trade argument that took place between several WTO Member States with regard to the import of genetically modified organisms (GMOs) between 2003 and 2006. It will demonstrate how the power relations of neoliberalism, sovereignty and biopolitics all coexist, albeit in dispersed, complex and indeed mutually contesting ways, in a specific regime of self-proclaimed "global governance". In revealing this complex relationship of different power relations, the case study will also challenge the prevalent (neo-) Marxist approach in the critical analysis of global economic practices, which subordinates an institution such as the WTO to general theories of power, hegemony, and ideology.

Questioning Universals: The Merits of a Foucauldian Approach to IR

My engagement with David Chandler's article will focus on two points that are highly problematic from a Foucauldian point of view: Chandler's lack of questioning "universals", such as democracy, politics, equality, and "the subject", from which he deduces the questions that he deems theoretically and analytically relevant, and the distinction he makes between "reality" and "fiction". In addition, I will also counter the allegation made by Jan Selby and Jonathan Joseph that Foucault does not provide a sufficient basis for paying tribute to the particularities of "the international". My argument will hopefully contribute to making a case for the merits of a Foucauldian approach to IR, before this approach is then applied to the case of EC Biotech in the second half of the article.

In his piece, Chandler compares the liberal-cosmopolitan critique of nation-state-based politics with the radical poststructuralist critique of cosmopolitan claims. My aim is not to prove as right or wrong his reading of cosmopolitans and poststructuralists. Instead, I want to show that Foucault's work itself (on which, according to Chandler, the poststructuralist critique is partly based)[2] both takes on board much of his criticism and reveals the implicit problems of his approach to poststructuralist work.

The linchpin of Chandler's article about the commonality of the cosmopolitan and the poststructuralist critique is that both supposedly reach beyond the national as the basis for political struggle without recognising that there is no

1. David Chandler, "Critiquing Liberal Cosmopolitanism? The Limits of the Biopolitical Approach", *International Political Sociology*, Vol. 3, No. 1 (2009), pp. 53–70.

2. *Ibid.*, p. 55.

"conceptual grounding" for a political community once the latter's connection to "citizenship" is broken. Following Chandler, both approaches consequently fail to recognise that contestations of political power are enabled precisely because of that link.[3] However, Foucault follows a very different epistemology that does not take any conceptual link as a pre-analytical given. Consequently, Foucauldians do not (just) reject the idea of a state-based political community but aim to reveal the particular configurations of power/knowledge involved in the formation of *any* political claim that leads to the establishment of political entities and the definition of certain questions as being decisive. If the difference in epistemology is taken as the linchpin of comparison, there seems to be more common ground between cosmopolitans and Chandler himself than between the former and poststructuralists: both take the link they establish (one between cosmopolitan rights and the existence of a global community, the other between legal representation and political community) for granted, while poststructuralists (should) question the essentialisation of both links. This relentless questioning does not imply that the concepts and entities that are developed and constituted based on these claims (subjects, communities, rights) are not real, as Chandler claims with regard to the poststructuralist critique.[4] From a Foucauldian point of view, the distinction between "real" and "fictitious" needs to be doubted; which does not mean that everything becomes fictitious, but that everything is real, as everything is "practice".[5] Consequently, both the practices of the nation-state and the practices that question the nation-state based on cosmopolitan rights are real—the so-called "Battle of Seattle" that disrupted the WTO trade negotiations in 1999 was no less real than political protests within a nation-state. Chandler links the "reality" of concepts such as "democracy" to the nation-state due to the legal formalisation that takes place at that site. In contrast, he declares the engagement with practices of global democracy as being based on wishful thinking and empirically non-existent.[6] What Chandler does not acknowledge is that his prioritisation of legal equality over (potential) substantial equality through equating the latter with the former is also a normative choice, which makes him become the object of his own critique.

With a Foucauldian perspective it is impossible to speak of the existence of "universals" and, closely linked to that, to establish a general theory of political power. Chandler claims that poststructuralists in IR suggest, with regard to political struggle, that "a new universal subject may be emerging from below", which finds itself in opposition to the one that is promoted by the cosmopolitans.[7] In contrast, for Foucault the interesting question is what emerges out of an analysis if its starting-point is not a "universal".[8] He avoids the paradox of needing a transhistorical, universal notion of the object of investigation that precedes its deconstruction by asking "how one might explore history" if the entity in question "did not

3. *Ibid.*, p. 56.

4. *Ibid.*, p. 61.

5. Michel Foucault, *The Essential Works of Michel Foucault 1945–1984*, Vol. III: *Power* (ed. J.D. Faubion) (London: Penguin, 1994), p. 225. Cf. Foucault's notion of "reality" in *ibid.*, p. 125, and to his understanding of "genealogy", in *ibid.*, pp. 118–119.

6. Cf. Chandler, *op. cit.*, p. 60.

7. *Ibid.*, p. 56.

8. Michel Foucault, *Security, Territory, Population: Lectures at the Collège de France, 1977–78* (Houndmills and New York: Palgrave Macmillan, 2007), p. 3.

always-already exist".[9] The result of such an investigation cannot be the affirmation of a "new" universal. Foucault's genealogical project was concerned about locating power relations in a complex and flexible constellation of diverse and mutually constituting as well as destabilising tactics and strategies; always at a particular site, at a particular time, through the analysis of particular practices.

This poses a problem for IR theorists, who in the past have mainly defined their task as finding general theoretical explanations of "the international", the independent existence of which has been taken for granted. Consequently, Jonathan Joseph questions the "utility" of Foucauldian concepts for IR, as they supposedly do not recognise its "distinctive properties and problems".[10] Jan Selby claims in a similar way that the "traditional" concerns of IR theory (war, inter-state relations, foreign policy, diplomacy, security) are "a long way" from Foucault's interests.[11] However, what both Joseph and Selby do not acknowledge is that Foucault's work was all about challenging the "traditional" concerns and assumptions of various disciplines (philosophy, psychology, medicine, etc.). It is correct that Foucauldian theorists in IR encounter difficulties if they are determined to maintain International Relations as something that can be distinguished from other social and political relations, and if they are reluctant to question its traditional objects of investigation. They then run the risk of making Foucault "fit" their approach, instead of questioning the ontological claims of the discipline itself.

The "Triangle" and the "*Dispositif*": Clarifying Foucauldian Concepts

In *Security, Territory, Population* Foucault is keen to depict the power relations he has engaged with in his genealogical work, namely sovereignty, discipline and governmental management, as a "triangle" in which the different modes of power intersect instead of substituting each other as singular manifestations of political rule.[12] By using the metaphor of the triangle, Foucault attempts to counter any conceptualisation of power in terms of different systems, and to analyse it instead as "heterogeneity" that implies "tensions, frictions, mutual incompatibilities, successful or failed adjustments, unstable mixtures, and so on".[13]

However, the metaphor of the triangle, as well as the concept of the strategy, has problematic connotations when it comes to this aim. Foucault's terminology seems to lend itself easily to a conceptualisation of power that, through codification and strategisation, moves bottom up in linear fashion from the micro- to the macro level. This leads Bob Jessop, who engages with Foucault's conceptualisation of the state, to implicitly place more emphasis on movements of consolation than on existing heterogeneities and internal tensions, which makes the state

9. Bob Jessop, "From Micro-powers to Governmentality: Foucault's Work on Statehood, State Formation, Statecraft and State Power", *Political Geography*, Vol. 26, No. 1 (2007), p. 36; referring to Foucault, *Security, Territory, Population*, *op. cit.*, p. 3.

10. Jonathan Joseph, "Neo-liberalism, Governmentality and Social Regulation", Paper prepared for the SAID Workshop, 17 April 2007.

11. Jan Selby, "Engaging Foucault: Discourse, Liberal Governance and the Limits of Foucauldian IR", *International Relations*, Vol. 21, No. 3 (2007), p. 331.

12. Foucault, *Security, Territory, Population*, *op. cit.*, p. 107.

13. Michel Foucault, *The Birth of Biopolitics: Lectures at the Collège de France, 1978–1979* (Houndmills and New York: Palgrave Macmillan, 2008), p. 21.

emerge at the end of a process of continuing codification and institutionalisation.[14] As I will show in the second part of this article, if a global perspective is taken, the state itself becomes a particular strategy that both constitutes and undermines regimes of global governance and vice versa—without "global governance" becoming the new end of the process. Equally, the triangle evokes the idea of governmental power as existing in a closed system that consists of fixed and stable axes of sovereignty, governmental management, and discipline.

Kimberly Hutchings claims that Foucault uses the vocabulary of power, strategy and tactics to point out the "perpetual openness which underlies apparently fixed and solid realities".[15] This claim is supported by one of the well-known passages in *The Will to Knowledge: The History of Sexuality, Volume 1*, in which Foucault argues that power must first of all be understood as the "the multiplicity of force relations immanent in the sphere in which they operate ... as the process which, through ceaseless struggles and confrontations, transforms, strengthens, or reverses them".[16]

However, a more thorough analysis of the chapter reveals that Foucault makes a distinction: on the one hand, there is power that is "everywhere", consisting of "unbalanced, heterogeneous, unstable, and tense force relations". On the other hand, there is "Power" (with capital P) that, "insofar as it is permanent, repetitious, inert and self-producing, is simply the over-all effect that emerges from all these mobilities". Power is the attempt to "arrest" power's mobility and to integrate it through codification.[17] A useful concept for understanding the relationship of power and Power is Foucault's *dispositif. Dispositifs* are formations of power relations that concentrate certain "effects" of power (= Power) in a specific space, but that can never be confined to it. Instead, they require "heterogeneity" and "diversity" (=power). The notion of the *dispositif* does not imply linear bottom-up or top-down strategisations but points at the "arrest" and integration of a heterogeneous set of power relations at a multiplicity of sites and a variety of scales.

It must be considered that Foucault sums up his famous elaborations on power under the title "Method".[18] Each of his books "is a way of dismantling an object, and of constructing a method of analysis toward that end".[19] In the case of his work on sexuality, and his later work on governmentality, the object is to dismantle the sovereign account of power as being unitary—so the appropriate (counter-) method is to analyse power as heterogeneous and multiple. Consequently, the "triangle" represents Foucault's constant attempt to forego a "theory of the state"; as does the "strategy" that he describes as the "logic of possible connections between disparate terms which remain disparate ... the logic of connections between the heterogeneous and not the logic of homogenisation [...]".[20] From a

14. Jessop, *op. cit.*, p. 39.

15. Kimberly Hutchings, "Foucault and International Relations Theory", in M. Lloyd and A. Thacker (eds.), *The Impact of Michel Foucault on the Social Sciences and Humanities* (Basingstoke and London: Macmillan; New York: St. Martin's Press, 1997), pp. 105, 125.

16. Michel Foucault, *The Will to Knowledge: The History of Sexuality, Volume 1* (London: Penguin, 1981), p. 92.

17. *Ibid.*, p. 93.

18. Foucault, *The Will to Knowledge*, *op. cit.*, p. 92.

19. Michel Foucault, *Remarks on Marx* (New York: Semiotext(e), 1991), p. 28.

20. Foucault, *The Birth of Biopolitics*, *op. cit.*, pp. 42, 76.

different point of view, namely that of complexity theory, John Urry argues in a similar way, with regard to the global sphere, that the "linear metaphor of scale" should be replaced by the "metaphor of connection", because any notion of macro, micro, etc. presumes that "there are entities with separate and distinct essences that are brought into external juxtaposition with each other".[21] From this perspective, the notion of the "triangle" is problematic, as it conveys the idea of closed geometrical space. Foucault's notion of the *dispositif* is far better to adequately represent his project.

I will now engage with three significant power paradigms that are used in contemporary critical and Foucauldian IR, and will show how they lose their universal and homogeneous status once they are confronted with a particular set of practices at a particular site.

Contesting Universal Paradigms I: Neoliberalism

According to David Harvey, neoliberalism is a theory of political economic practices whose components are few and easy to determine. Neoliberalism is about "liberating individual entrepreneurial freedoms and skills within an institutional framework characterized by strong private property rights, free markets, and free trade".[22] In most critical accounts, regimes of global governance are equated with the worldwide implementation of this particular politico-economic project.[23] The International Monetary Fund (IMF), the World Bank and the WTO are considered key organisations through which the neoliberal agenda "attempts to install itself as the only, the necessary and most desirable way".[24] Resistance to this policy is conceptualised according to respective theoretical perspectives and political beliefs: communitarians perceive either the individual nation-state or regional communities as the right location for decision making;[25] world-system theorists are enthusiastic about the potential of the "democratic socialist regimes" in the Southern semi-periphery;[26] and cosmopolitans locate resistance in the hundreds of civil initiatives that are part of the "global civil society".[27]

Many of these accounts are embedded in Marxist and neo-Marxist theoretical traditions.[28] This is not a surprise, considering that the Marxist belief in the necessity and possibility of overhauling capitalism to bring liberation and emancipation for humanity provides both critical theorists and political activists with a strong

21. John Urry, *Global Complexity* (Cambridge: Polity Press; Oxford and Malden, MA: Blackwell Publishing, 2003), p. 122.

22. David Harvey, *A Brief History of Neoliberalism* (New York: Oxford University Press, 2005), p. 2.

23. See, for example, Naomi Klein, *The Shock-doctrine: The Rise of Disaster Capitalism* (London: Penguin, 2008).

24. John Clarke, "Dissolving the Public Realm? The Logics and Limits of Neo-liberalism", *Journal of Social Policy*, Vol. 33, No. 1 (2005), p. 30.

25. See, for example, Alejandro Colás, "The Power of Representation: Democratic Politics and Global Governance", *Review of International Studies* (special issue), Vol. 29 (2003), p. 98.

26. See, for example, Christopher Chase-Dunn and Barry Gills, "Waves of Globalization and Resistance in the Capitalist World-system: Social Movements and Critical Global Studies", in R.P. Appelbaum and W.I. Robinson (eds.), *Critical Globalization Studies* (New York and Abingdon: Routledge), p. 52.

27. See, for example, Mary Kaldor, "'Civilising' Globalisation? The Implications of the 'Battle in Seattle'", *Millennium*, Vol. 29, No. 1 (2000), pp. 105–114.

28. Wendy Larner, "Neoliberalism?", *Environment and Planning D: Society and Space*, Vol. 21, No. 5 (2003), p. 511.

and inspiring theoretical base. Presenting neoliberalism as a "coherent ideological project with clear and unambiguous origins, whose spread is sustained and circulated by an identifiable set of institutions",[29] which are grounded in capitalist relations, supports political struggle (which arguably is the political aim of critical research) by outlining a clear-cut enemy. However, interpreting global power relations in this way has its price, as it excludes the possibility "to think about the multiple forms that political strategies, techniques, and subjects take".[30] If political resistance is based on the forcing together of "phenomena and situations that are not necessarily similar or comparable" to sustain the idea of hegemony, and on the glossing over of heterogeneities and divergences in the exercise of power, it might not gain the necessary traction. It is exactly the rejection of such a move, based on an acknowledgement of the heterogeneity of political practices at specific political sites, which makes a Foucauldian approach significant.

In recent years, a number of students have attempted to reconcile Marxist conceptualisations of neoliberalism with the analysis of "advanced liberalism" in Foucauldian "governmentality studies".[31] However, according to Clive Barnett, this attempt merely leads to the application of Foucault's work for the purpose of "shoring up the holes" in Marxist theory.[32] Foucault's concepts of "discourse" and "(neoliberal) governmentality" are used to explain how power is deployed "to achieve dominance" in the global sphere;[33] in other words, they are regarded as strategies that operate on the surface of an underlying general structure. Jonathan Joseph's (affirming) use of Foucault's concept of governmentality for a study of neoliberalism is a good example of such an appropriation. In contrast to Foucault, who does not provide any ground for an ontology outside of the practices that constitute reality in multiple ways, Joseph argues that "governmentality itself is dependent on underlying social conditions", which can be adequately described as hegemonic.[34]

However, Joseph rightly points out that "the narrow view of neo-liberalism as deregulated free market capitalism" dismisses the fact that neoliberalism is itself a form of social regulation. His emphasis on neoliberalism as a project of (governmental) regulation and not of *de*regulation elucidates one of the central features of Foucault's concept of governmentality. Foucault describes liberalism as a "principle and method of the rationalization of the exercise of government" which obeys "the internal rule of maximum economy".[35] "Government" and "liberalism" have to be regarded as existing in an inseparable relationship. This does not deny the fact that the primary imperative of liberalism as an "art of government" is the *limitation* of government—but this limitation is achieved through governmental planning and regulation itself. Any analysis of liberalism should therefore not focus on its discourse of deregulation and laisser-faire but on how it attempts to limit itself *through* regulation. This insight is even more significant for the analysis of *neo*liberalism, because one of the central differences between liberal and neoliberal practices is the latter's representation of laisser-faire as

29. Clive Barnett, "The Consolations of 'Neoliberalism'", *Geoforum*, Vol. 36 (2005), p. 8.
30. Larner, *op. cit.*, p. 512.
31. *Ibid.*, p. 511. See also Barnett, *op. cit.*, p. 8.
32. Barnett, *op. cit.*, p. 8.
33. See, for example, Clarke, *op. cit.*, p. 30. Cf. Barnett, *op. cit.*, p. 9.
34. Joseph, *op. cit.*, p. 12.
35. Foucault, *The Birth of Biopolitics*, *op. cit.*, p. 318.

"naïve naturalism". Foucault maintains that neoliberalism is profoundly constructivist—it is based on the conviction that the economic logic of liberalism, centred on the processes of the market, is not natural. Instead, it can work only under certain conditions which have to be "carefully and artificially constructed". This presupposes an "indefinitely active policy".[36]

The WTO is an excellent example of the necessity of active governance in neoliberal logic. Following the Marrakesh Declaration of 1994, the organisation has been established to support the liberalisation of global trade through the construction of a "stronger and clearer framework".[37] The trade rules that had already been agreed on in the framework of the General Agreement on Tariffs and Trade in 1947 (GATT) were supposed to be in need of institutionalisation for the sake of more "fair[ness]".[38] In an affirming statement, the then-Managing Director of Operations, Policies and Programmes of the World Bank, Gautam S. Kaji, uses an interesting metaphor to describe the role of the WTO, the World Bank and the IMF in international trade: he says their central role is the setting up of a "single-level playing field" of trade.[39] Underlying this understanding is the idea that the benefit of the whole is not something that can be achieved through detailed control, because the processes of economies and societies are too obscure for that. Therefore, members of the WTO, IMF and World Bank are supposed to reject individual preference-seeking through mechanisms of control, and favour the establishment of a *game* in which they all participate. The metaphor of the game elucidates that the society that needs to be regulated by reference to the market should be a "society in which the regulatory principle should not be so much the exchange of commodities as the mechanisms of competition".[40] For this, the "stage" needs to be set.[41] This implies following tasks for the involved international institutions: (1) establishing the rules of the game through general agreements, and checking that they are kept. In the case of the WTO, this implies particular attention to the rule of non-discrimination between the "players" of the game, and the securing of their formal equity. (2) Calling for a certain equipping of the players that prevents them from falling under a determined threshold under which they are not capable of playing any more. For example, in 1996, the then-Director-General of the WTO, Renato Ruggiero, asked the governments of the developed world to reduce the debts of development countries to bring them into a "sustainable debt position".[42] It is important to notice that a *reduction* does not mean an *annulment*. It is not the aim of the WTO to achieve substantial equity between the players.

36. *Ibid.*, p. 120.

37. World Trade Organization, *Marrakesh Declaration of 15 April 1994*, Article 1, available: <www.wto.org/english/docs_e/legal_e/marrakesh_decl_e.pdf> (accessed 27 April 2009).

38. *Ibid.*, Article 2.

39. World Trade Organization, Ministerial Conference, Singapore, 9–13 December 1996: Statement by Mr Gautam S. Kaji of the World Bank, available: <http://www.wto.org/english/theWTO_e/minist_e/min96_e/st33.htm> (accessed 27 April 2009).

40. Foucault, *The Birth of Biopolitics, op. cit.*, p. 147.

41. World Trade Organization, Ministerial Conference, Singapore, 9–13 December 1996: Statement by Mr Michel Camdessus, International Monetary Fund, available: <http://www.wto.org/english/theWTO_e/minist_e/min96_e/st30.htm> (accessed 27 April 2009).

42. World Trade Organization, Address by Renato Ruggiero, Director-General, World Trade Organization, 1996, available: <http://www.wto.org/english/theWTO_e/minist_e/min96_e/sing_dg_e.htm> (accessed 27 April 2009).

The techniques through which the objectives of the international institutions are to be achieved are formal de-hierarchisation in negotiations (=formal equity), the securing of policy coherence, the formalisation of cooperation, and institutionalisation. This resembles what Foucault calls an element of "American neoliberalism": the society that is based on exclusionary mechanisms applied to those who cannot be normalised is replaced with a conceptualisation of society in which there is an "optimization of systems of difference", in which the "field is left open to fluctuating processes" and in which "action is brought to bear on the rules of the game rather than on the players".[43] It is the kind of neoliberalism that David Harvey and Jonathan Joseph define as Neoliberalism as such.

However, the case of EC Biotech elucidates that there is more than one neoliberalism employed in the attempt to govern international trade. In 2003, the United States, Canada and Argentina launched a trade dispute case against the EC at the WTO with regard to the latter's "de facto moratorium" on the import of new GMOs. Owing to immense public opposition to the first GMOs that entered the EC between 1996 and 2000, EC Member States had blocked approval authorisation from 1998 onwards.[44] Half a year after the US–Canadian–Argentinean complaint at the WTO, the European Council of Ministers and the European Parliament passed regulations on labelling and tracking GM crops. This followed a revision of the EC Directive that deals with the so-called "deliberate release of GMOs into the environment" in 2001. As a consequence, the moratorium was lifted.[45] The WTO panel's decision on the case in 2006 confirmed the illegality of the former moratorium—a decision which the EC regarded as being largely of historical interest owing to its change of policies.[46]

The EC position was embraced by anti-GM non-governmental organisations (NGOs) and also by critics of the unlimited liberalisation of trade. However, I will argue that, in this particular case, the EC position is at least as neoliberal as the position of its accusers. What is different between the EC and the US–Canadian–Argentinean position is that the former attempts to construct a market for GMOs that enables the consumer to choose between GM and non-GM products, while the latter deem that superfluous.

In their approach to genetic engineering, the United States, Canada, and Argentina are primarily concerned about the promotion of societal progress through the means of technology. This is based on the underlying assumption that technological development generally leads to an increase in human well-being and should therefore be promoted out of principle. In EC Biotech, the United States, for example, demands from the EC that it can prove on a case-by-case basis the danger of each specific GMO that is not approved, along the lines of the WTO agreements.[47] When it comes to the benefits of biotechnology,

43. Foucault, *The Birth of Biopolitics*, *op. cit.*, pp. 259–260.

44. Helge Torgersen *et al.*, "Promise, Problems and Proxies: Twenty-five Years of Debate and Regulation in Europe", in M.W. Bauer and G. Gaskell (eds.), *Biotechnology—The Making of a Global Controversy* (Cambridge: Cambridge University Press, 2002), pp. 60, 73.

45. Charan Devereaux, Robert Z. Lawrence and Michael D. Watkins, *Case Studies in US Trade Negotiations*, Vol. II: *Resolving Disputes* (Washington, DC: Institute for International Economics, 2006), p. 329.

46. *Ibid.*, p. 336.

47. World Trade Organization, *European Communities—Measures Affecting the Approval and Marketing of Biotech Products: Reports of the Panel [WT/DS291/R, WT/DS292/R, WT/DS293/R]* (2006), available: <http://www.wto.org/english/tratop_e/dispu_e/cases_e/ds291_e.htm> (accessed 3 November 2008), 4.133.

the United States remains very general, mainly drawing on its "potential" and the benefits it brings "in principle".[48] The approach is based on some of the "unstated assumptions of 'modern' (and particularly western) societies: material growth, [and] the power and efficacy of scientific reason".[49] This also implies the non-toleration of a "zero risk" approach in policy—the United States stresses that it is necessary to create, in contrast, "an appropriate level of protection".[50] It argues along the lines of the WTO Sanitary and Phytosanitary (SPS) Agreement, which explains under which conditions a WTO Member can ban a product from being imported with the claim that it causes harm to its population, plants, and animals. In Article 5 of the Agreement it is stressed that for defining an appropriate level of sanitary and phytosanitary protection it is necessary to take into account the "character of human health risks to which people voluntarily expose themselves", implying that there is a normal level of risk that needs to be tolerated.[51] From a Foucauldian point of view, this approach can be described as traditional biopolitics in the sense that action is based on the idea that what is "risky and inconvenient . . . will never be completely suppressed".[52]

For the EC, governmental action is guided by the principle of the avoidance of harm or adverse effects on human health and the environment. It is this "prudent and precautionary" approach which is prevalent in EC Directive 2001/18/EC and in its argument at the WTO.[53] The precautionary approach in environmental politics implies that "policy makers should act in advance of scientific certainty to protect the environment from incurring harm". It demands that "humans take care for themselves, their descendants and for the life-preserving processes that nurture their existence".[54]

However, this different biopolitical understanding to be found in the EC does not imply a complete ban of GMOs due to the potential irreversible effects they might have on the environment and human well-being. What the EC is really concerned about is the establishment of the right framework for the markets through which biotechnological products circulate. In terms of the governance of GMOs, the decisive difference between the United States, Argentina and Canada, and the EC is how they transform (or do not transform) GMOs in an object of governance. As Javier Lezaun shows in his excellent study of the EC approach to GMOs, the EC manages to "make the category of 'GM' real in the world by imposing increasingly stringent testing, identification and labelling

48. *Ibid.*, 4.136–4.142.

49. Andrew J. Jordan and Timothy O'Riordan, "The Precautionary Principle in Contemporary Environmental Policy and Politics", Paper prepared for the Wingspread Conference on "Implementing the Precautionary Principle", 23–25 January 1998, Racine, Wisconsin, available: <http://www.johnsonfdn.org/conferences/precautionary/jord.html> (accessed 23 January 2009).

50. World Trade Organization, *WT/DS291[2, 3]/R*, *op. cit.*, 4.177.

51. World Trade Organization, *Agreement on Sanitary and Phytosanitary Measures [SPS-Agreement]* (1995), Article 5.5, available: <http://www.wto.int/english/tratop_e/sps_e/spsagr_e.htm> (accessed 3 November 2008).

52. Foucault, *Security, Territory, Population*, *op. cit.*, p. 19.

53. European Communities, *Directive 2001/18/EC of the European Parliament and of the Council* (2001), Preamble, available: <http://eurlex.europa.eu/LexUriServ/LexUriServ.do?uri=OJ:L:2001:106:0001:0038:EN:PDF> (accessed 3 November 2008); World Trade Organization, *WT/DS291[2, 3]/R*, *op. cit.*, 4.334 (see also 4.332).

54. Jordan and O'Riordan, *op. cit.*

obligations on genetically engineered products".[55] In this way the EC effectively abandons the principle of substantial equivalence that the accusing parties in EC Biotech still employ. Canada, for example, argues that the physical properties of biotech and non-biotech canola/oilseed rape are "in all essential aspects, virtually identical". As they are also "intended to be used interchangeably", Canada concludes that both products "must be considered 'like products'".[56] In contrast, the EC assumes that there is indeed a substantial difference and therefore defines its governmental task as making it absolutely transparent to the market participants whether or not a product is genetically modified, both through tracking/labelling regulations and, in agriculture, through guaranteeing the "coexistence" of transgenic, conventional and organic agriculture.[57] The aim is to extend the choices that producers and consumers have on the market, and the EC is thereby in line with what Mitchell Dean calls "advanced liberal government": individuals and communities become responsible for managing their risks themselves, which makes risk a feature of choice.[58]

The EC approach is part of a particular strand in neoliberalism that many critical accounts disregard: the neoliberal critique of neoclassical equilibrium models; a critique that makes use of the "complex models of self-organization" proposed by Friedrich von Hayek.[59] Hayek (and his supporters) emphasise the role of innovation in the advance of society, but also the impossibility of knowing which innovation will make society progress in advance. Both market and society are regarded as constantly "evolving" in non-linear fashion, without the possibility of predicting the direction these evolutions take (thereby contesting one of the basic assumptions of neoclassical economics). As a consequence, the so-called market "discovery-element" needs to be respected by any regulating body. The "social value of innovation" can never be predicted "ex ante except that it extends the choice set, leaving their non-acceptance as one alternative which has to be respected by a theoretical observer".[60]

The similarities between this neoliberal critique and the EC approach to biotechnology are striking. According to more traditional neoliberal and ordoliberal economic theorists, "the state has to define the rules of the economic game, while the choice of actual moves in the game would be up to economic agents". In contrast, from a perspective that is often called "evolutionary economics", setting the rules should be regarded as a "move in the game itself".[61] The EC Member States' blocking of the import of GMOs until a reversal of EC policy enabled the tracing and labelling of GMOs can be characterised as such a move. It was followed by the construction of a market for GMOs that would extend choices.

55. Javier Lezaun, "Creating a New Object of Government: Making Genetically Modified Organisms Traceable", *Social Studies of Science*, Vol. 36, No. 4 (2006), p. 500.

56. World Trade Organization, *WT/DS291[2, 3]/R*, *op. cit.*, 4.225–4.226.

57. Cf Lezaun, *op. cit.*, p. 1.

58. Mitchell Dean, *Governmentality: Power and Rule in Modern Society* (London, Thousand Oaks and New Delhi: Sage, 1999), p. 166.

59. Melinda Cooper, *Life as Surplus: Biotechnology and Capitalism in the Neoliberal Era* (Seattle: University of Washington Press, 2008), p. 10.

60. Gerhard Wegner, "Reconciling Evolutionary Economics with Liberalism", in K. Dopfer (ed.), *Economics, Evolution and the State: The Governance of Complexity* (Cheltenham and Northampton: Edward Elgar, 2005), p. 62.

61. Matthias Klaes, "Historical Economics and Evolutionary Economic Policy—Coasean Perspectives", in Dopfer (ed.), *op. cit.*, p. 92.

The dispute case at the WTO makes clear how controversial this move was in the eyes of the EC trade partners, which elucidates why neoliberalism cannot be defined as a singular hegemonic project. Instead, one can recognise a multiplicity of different neoliberal practices at different institutional sites. In contrast to David Harvey, I would argue that neoliberalism does not simply adhere to the "free market principles of neoclassical economics"[62] but, in some expressions, also criticises these principles *from* a neoliberal perspective.

Determining neoliberalism's overall global structure, as Harvey attempts to do, can only fail because it steamrolls the complexities involved in order to gain a smooth and linear narrative. Heterogeneity does not start with the distinction between "liberal" and "illiberal" practices, as Jonathan Joseph argues when he claims that the framework of governmentality can be applied to societies of advanced liberalism, but not to others.[63] It starts much earlier, namely within the paradigm of (neo-)liberalism itself. Joseph claims that Foucault is of little help in "analys[ing] the illiberal, uneven, and irreducible character of much that goes on within international relations".[64] However, the more the contemporary exercise of political power is decentred, the more uneven and irreducible "the international" becomes; as indeed does any other site of analysis.

Therefore, Foucauldian IR theorists should focus on the various dimensions of particular practices at particular local sites at a particular time. Such an analysis can turn, in a typical Foucauldian manner, traditional assumptions, concepts and theories upside down—in this case the assumption that the United States and the WTO are key forces in the global spread of the neoliberal agenda per se.

Contesting Universal Paradigms II: Sovereignty

For Foucault, using the framework of governmentality implies studying the rationalisation of government "in the exercise of sovereignty".[65] Although Foucault constantly challenges sovereignty as the essential modern paradigm, it still provides the ultimate framework for his analyses. Giorgio Agamben alleges that Foucault never managed to bring together what Agamben calls the "double bind" of the techniques of individualisation and (sovereign) totalisation that are involved in the project of governmentality.[66] Agamben locates the "unitary center" for this connection in what he conceptualises as the original "secret tie" between sovereignty and biopolitical life. Following him, both are mutually constituted in their inclusion into the realm of "law" through the exclusion of "bare life" in the sovereign "exception".[67]

In recent years, Foucauldian IR theory has been increasingly inspired by Agamben's work, due to the introduction of a rising number of anti-liberal security measures in Western liberal states in the aftermath of 9/11. His synthesis of the Foucauldian concept of biopolitics with sovereign power seems to provide a very useful basis for theorists in search of an explanation for the sudden rise of

62. Harvey, *op. cit.*, p. 20.
63. Joseph, *op. cit.*, p. 13.
64. *Ibid.*
65. Foucault, *Security, Territory, Population*, *op. cit.*, p. 2.
66. Giorgio Agamben, *Homo Sacer: Sovereign Power and Bare Life* (Stanford: Stanford University Press, 1998), pp. 15–16.
67. *Ibid.*, pp. 6–7.

sovereign practices.[68] A good example is the camp in Guantanamo Bay, whose inmates are indeed reduced to "bare life" and placed in an infinite "state of exception"—in an age that, following Foucault, is characterised by the promotion of life.

However, Agamben neglects Foucault's constant attempt to disperse what appears uniform. Following Andrew Neal, Agamben makes the mistake of reading Foucault structurally instead of genealogically, which results in a "philosophical totalization" of sovereignty.[69] Instead of assuming that the universal of sovereignty does not exist, Agamben takes it as a given, instead of decentring, his whole aim is to find the centre, and instead of looking for immanent discontinuities, Agamben is determined to trace the continuities in the metaphysical discourse of sovereign exceptionality.[70] Michael Dillon explains this difference by referring to the theorists' different analytical standpoints: The philosopher Agamben "takes Foucault's account of biopolitics away from history and relocates it back in the centre of [the] key determinants of political philosophy"; in other words, he "ontologises" the concept of biopolitics.[71] In contrast, Foucault is more interested in giving a historical account of the emergence of biopolitics as a specific technology and rationale in the government of modern society. He acknowledges the complementary relationship of sovereignty and biopolitics by including sovereignty as one important set of power relations in his governmental "triangle". However, in contrast to Agamben's account, both represent different "zones of power" that are characterised by an "irreconcilable tension", although they constantly "intersect" in governmental practice.[72]

This contribution aims to elaborate on what this means in practice. An empirical investigation has the advantage of being able to show how different sets of power relations can be seen as "interwoven in ways not made evident by their theoretical articulation".[73] Relating the analysis to Agamben's work, I will engage with practices of "exceptionalism" in the context of the international governance of trade. This engagement requires extending the common understanding of the exception, which is too often limited to state-internal emergency measures in cases of external threat and internal civil unrest. In his history of the state of exception, Agamben himself points out that in France, after the First World War, the exception left the sphere of *military* emergency and expanded into the sphere of *economic* emergency, when the French administration asked for "full power" over financial matters.[74] Equally, the American New Deal gave the president "an unlimited power to regulate and control every aspect of the economic life of the country".[75] However, while the state of exception in democratic countries

68. See, for example, the contributions in Jenny Edkins, Véronique Pin-Fat and Michael J. Shapiro (eds.), *Sovereign Lives: Power in Global Politics* (Abingdon and New York: Routledge, 2004).

69. Andrew Neal, "Foucault in Guantánamo: Towards an Archaeology of the Exception", *Security Dialogue*, Vol. 37, No. 1 (2006), pp. 32, 39.

70. Agamben, *op. cit.*, pp. 8, 11.

71. Michael Dillon, "Cared to Death: The Biopoliticised Time of Your Life", *Foucault Studies*, No. 2 (2005), p. 38.

72. Mika Ojakangas, "Impossible Dialogue on Bio-power: Agamben and Foucault", *Foucault Studies*, No. 2 (2005), p. 26.

73. Miguel de Larrinaga and Marc G. Doucet, "Sovereign Power and the Biopolitics of Human Security", *Security Dialogue*, Vol. 39, No. 5 (2008), p. 519.

74. Giorgio Agamben, *State of Exception* (Chicago and London: University of Chicago Press, 2005), p. 13.

75. *Ibid.*, p. 22.

historically implied the transfer of "full powers" to the executive, in the context of certain practices of international governance it is about the exceptional transfer of powers that have been given to international institutions back to the sovereign state. Moreover, what is suspended is not a legal constitution but the (liberal) freedom of circulation.

The exception, defined as the monopoly over the decision to create and suspend the rule, is a defining feature of sovereignty in all areas of governance. However, rather than looking for the relationship between sovereignty and *juridical rule*, which underlies the traditional interest in the exception,[76] this case requires the analysis of the link between sovereignty and Foucault's notion of the *biopolitical norm*.[77] In EC Biotech it can be found in the sovereign capacity to limit the (liberal) freedom of circulation through the suspension of certain circulations that are considered "bad" for the (biopolitical) well-being of the population.[78] The case makes clear how the WTO both constitutes and constrains this expression of sovereignty through the attempt to integrate the biopolitical exception into its own procedures.

In EC Biotech, the United States, Canada, and Argentina accuse both the EC and individual EC Member States of not complying with WTO Agreements. According to their allegation, the blocking of the import approval of GMOs from a majority of the EC Member States led to a general suspension of all approval procedures.[79] This implied an unjustifiable and arbitrary prohibition of the import of GMOs, as then-EC Commissioner for Agriculture Margot Wallström pointed out when she argued that there was an "arbitrary line" drawn to stop all approval, because it was not related to any scientific evidence.[80]

If an action is sovereign, it does not need external justification. This is also stressed in a so-called Amicus Curiae Submission that several anti-GM NGOs contributed to the WTO case to support the EC position. They argue that the EC moratorium represents an "expression of political intent by autonomous sovereign states". Following this argument, the moratorium is "not a matter for WTO scrutiny", as this would imply an "inappropriate interference with the political process internal to the EC".[81] Political intentions and procedures are located within the realm of sovereignty, which is contrasted to the realm of regulatory global governance. While the latter enables the freedom of circulation, the former has the capacity to suspend it. A moratorium, as an act of ultimate suspension, is firmly put into the zone of sovereign politics by all parties.

The decisive difference between the traditional sovereign decision on the exception, and the relationship of sovereignty and the exception in the case of the WTO, is that any sovereign suspension of circulation is accepted by the international community of states only if it is temporary and provisional. If a sovereign state

76. *Ibid.*, p. 1.

77. Following François Ewald, the (biopolitical) "norm" is opposed to the "juridical". François Ewald, "Norms, Discipline, and the Law", *Representations* (special issue), Vol. 30 (1990), p. 138.

78. Cf. Foucault, *Security, Territory, Population*, *op. cit.*, p. 18.

79. World Trade Organization, *WT/DS291[2, 3]/R*, *op. cit.*, e.g. 4.132–4.133.

80. *Ibid.*, 4.147.

81. Alice Palmer (on behalf of 15 NGOs), *Request for Permission to Submit Information to the Panel by the Following Non-Parties (Amicus Curiae Submission)* (2004), 3.1.1 (80), available: <http://www.genewatch.org/uploads/f03c6d66a9b354535738483c1c3d49e4/PublicInterestAmicus_2.pdf> (accessed 22 January 2009).

wants to belong to this community, it is forced to justify its action, with this justification being accepted and finally integrated into the norm. The unlimited exercise of sovereignty cannot be tolerated among WTO Members, because it leads to a standstill of circulation; a "freezing"; a slow-down that is inefficient and a "hindrance to progress". It disrupts the continuity of movement and can therefore be accepted only from the perspective of liberal governance if its appropriateness can be justified and if it is not "excessive".[82]

One of the WTO Agreements that specifies the conditions for the legitimate temporary suspension of the freedom of circulation is the SPS Agreement. This Agreement allows WTO Member States to protect their territory from "risks arising from the entry, establishment or spread of pests, diseases, disease-carrying organisms or disease-causing organisms" through an import ban, if the product to be blocked endangers the Member State's "human, animal, and plant health".[83] The distinction between good and bad circulation in this Agreement is therefore based on biopolitical grounds, and it is in this space that sovereignty intersects with the biopolitical imperative "to make life live" in governmental practice.

However, according to the US argument in EC Biotech, the central objective of the SPS Agreement is *not* the permitting of import bans under certain conditions. Instead it is, with reference to the Preamble of the Agreement, the minimisation of the negative effects of sanitary and phytosanitary regulation on trade. Put in this way, the end of the SPS Agreement is finally a liberal one and thereby part of the general aim of the WTO, which is the reduction of restrictions on international trade under the principle of non-discrimination. Following this argument, the SPS Agreement is not what could be called a biopolitical *concession* to liberal trade laws but rather a *constraint*, which means an attempt by the WTO to restrict the scope of sovereign action that is biopolitically justified.[84]

The integration of the biopolitical "exception" into WTO norms is based on what Foucault calls the "disciplinarisation of knowledge" in *Society Must be Defended*. The end of this type of discipline is the qualification (and disqualification) of knowledges through the establishment of a scientific community at the site of the university. There, knowledges are redistributed, communicated, and hierarchicalised within an overall field—science.[85] Beside the university, Foucault emphasises the role of the state in this process.[86]

In the WTO, the dominant knowledge on which decisions are supposed to be based is also science. Both state governments and international institutions have contributed to the selection, homogenisation and hierarchicalisation of qualified scientific knowledge through the institutionalisation of the "scientific community", which makes knowledge accessible and manageable. For example, the SPS Agreement refers to the "Codex Alimentarius Commission" for food security, the "International Office of Epizootics" for animal health, and the "International Plant Protection Convention" for plant health.[87] To determine the risk of individual products, the SPS Agreement demands a risk assessment that follows the

82. World Trade Organization, *WT/DS291[2, 3]/R*, *op. cit.*, 4.156, 4.168.

83. World Trade Organization, *SPS Agreement*, *op. cit.*, Annex A 1(a).

84. World Trade Organization, *WT/DS291[2, 3]/R*, *op. cit.*, 4.162.

85. Michel Foucault, *Society Must be Defended: Lectures at the Collège de France, 1975–76* (London: Penguin, 2005), pp. 183–185.

86. *Ibid.*, p. 186.

87. World Trade Organization, *SPS Agreement*, *op. cit.*, Annex A (3).

international standards that have been established by these organisations. The United States, Canada and Argentina argue that the safety of GMOs has been proven by these and other national and international scientific institutions.[88]

But the attempt to integrate the exception into the norm is ultimately not successful. The US argument does not acknowledge the ambiguity of the SPS Agreement Preamble, which names three aims:

> (...) *Desiring* to improve the human health, animal health and phytosanitary situation in all Members ... *Desiring* the establishment of a multilateral framework of rules and disciplines to guide the development, adoption and enforcement of sanitary and phytosanitary measures in order to minimize their negative effects on trade ... *Desiring* to further the use of harmonized sanitary and phytosanitary measures between Members.[89]

The objective of the Agreement is therefore not only the production of freedom for the circulation of people and goods but also the biopolitical improvement of human, animal and plant health that justifies sovereign action. The SPS Agreement attempts to integrate the exception, but is not able to reconcile its essential tension with the freedom of circulation.

Interestingly, this analysis has now arrived at a similar result to Giorgio Agamben's, but through the engagement with a particular set of practices: in this specific case of international liberal governance there is a bond created between sovereignty and biopolitics that is established through the presumed necessity of sovereign action on biopolitical grounds. But this bond is an immanent one that is embedded in a particular historical context. My analysis cannot comment on the existence of a potential transcendent "original" relationship that is based on the universal constitution of sovereignty through the inclusion/exclusion of "bare life".

Contesting Universal Paradigms III: Biopolitics

In another strand in contemporary Foucauldian IR, students have moved beyond the discipline's traditional interests, and engage with fields of governance that are "somehow neglected".[90] These are, for example, the international politics of infectious diseases, insurance, and the securitisation of global aviation; analysed from a perspective that they call the "biopolitics of security".[91] Instead of depicting biopolitics as a universal phenomenon, these theorists generally define it as a "specific historical ... evolving economy of power relations".[92] It is their

88. World Trade Organization, *WT/DS291[2, 3]/R*, *op. cit.*, 4.137–4.141.

89. World Trade Organization, *SPS Agreement*, *op. cit.*, Preamble; emphasis in original.

90. Michael Dillon and Luis Lobo-Guerrero, "Biopolitics of Security in the 21st Century: An Introduction", *Review of International Studies*, Vol. 34, No. 2 (2008), p. 265.

91. See, for example, Stefan Elbe, "Risking Lives: AIDS, Security and Three Concepts of Risk", *Security Dialogue*, Vol. 39, Nos. 2–3 (2008), pp. 177–198; Luis Lobo-Guerrero, "Biopolitics of Specialised Risk: An Analysis of Kidnap and Ransom Insurance", *Security Dialogue*, Vol. 38, No. 3 (2007), pp. 315–334; Mark Salter, "Imagining Numbers: Risk, Quantification, and Aviation Security", *Security Dialogue*, Vol. 39, Nos. 2–3 (2008), pp. 243–266.

92. Dillon and Lobo-Guerrero, *op. cit.*, p. 267.

conviction that Foucault's (biopolitical) "*dispositif* of security" has more to do with the circulation, cultivation, promotion and fructification of life than with "restraining and suspending".[93] Moreover, Michael Dillon, for example, acknowledges that there is not only one "securitizing game" going on in Western countries but many heterogeneous ones which "transact" and are also in conflict with each other.[94]

Still, in most of the associated analyses, the heterogeneity of biopolitics itself is neglected. A good example is Michael Dillon's (and Julian Reid's) argument about the global dominance of what he/they call(s) "recombinant biopolitics". Dillon contends that "the contingent" has become the "primary strategic principle" of securing life globally. This change in strategy is based on the acknowledgement that the "biological being" is no longer conceptualised in "simple linear fashion".[95] I will show, however, that in the case of EC Biotech the dominant conceptualisation of life is still based on this latter understanding; which once again reveals the complexities of power.

In EC Biotech the accusing parties' understanding of what happens in genetic modification is based on what I call a *linear-static* understanding of life processes, while the EC follows a *complex-dynamic* scientific approach. The WTO Dispute Panel decides to follow the linear-static approach in its ruling when it declares that the SPS Agreement that is based on the traditional understanding of life is applicable in this case.

According to the EC argument, the approach used by the accusing parties is one based on "mainstream scientific opinions".[96] The underlying scientific model is characterised by a notion of risk that locates danger at the level of the (static and stable) characteristics of a product and its direct linear causal-chain interaction with its immediate environment. It is generally characterised by a belief in the ultimate controllability of human action. Canada, for example, argues that "it is not the process through which a plant with novel traits is developed that determines the risk, but rather the characteristics of the inserted gene(s) and the host plant, the environment in which the plant is released and the use to which the plant is put".[97]

Brian Wynn points out that in genetics, "complexity as the limits of predictability ... is continually encountered", but at the same time denied.[98] The possibility of genetic modification is celebrated as a way of "increasing knowledge, and as a new dawn of precision-biotechnology", while the unpredictability of what happens in the process of modification and the lack of control of its consequences is only admitted "as a temporary, strictly limited problem".[99] According to Wynn, this is because the knowledge about the complexity dimensions of genetic life

93. Michael Dillon and Andrew Neal, "Introduction", in M. Dillon and A. Neal (eds.), *Foucault on Politics, Security and War* (Basingstoke and New York: Palgrave Macmillan, 2008), p. 11.

94. Michael Dillon, "Underwriting Security", *Security Dialogue*, Vol. 39, Nos. 2–3 (2008), pp. 309–332 (p. 326).

95. *Idem*, "Governing through Contingency: The Security of Biopolitical Governance", *Political Geography*, Vol. 26, No. 1 (2007), p. 46; *idem*, "Underwriting Security", *op. cit.*, p. 312. This argument is taken up again in Michael Dillon and Julian Reid, *The Liberal Way of War: Killing to Make Life Live* (New York and Abingdon: Routledge).

96. World Trade Organization, *WT/DS291[2, 3]/R*, *op. cit.*, 4.376.

97. *Ibid.*, 4.198.

98. Brian Wynn, "Reflexing Complexity: Post-genomic Knowledge and Reductionist Returns in Public Science", *Theory, Culture & Society*, Vol. 22, No. 67 (2005), p. 69.

99. *Ibid.*, p. 71.

processes is difficult to combine with assumptions or promises of "prediction-and-control".[100] What is denied is the complex gene–environment interaction that leads to the production of specific phenotypes. Following mainstream genetics, it is purely the genome that shapes the phenotype in a linear, deterministic, controlled way—the "arrows point strictly one-way from DNA to RNA to protein", and never the other way round.[101] Wynn concludes that biology is still embedded in the "epistemic culture of instrumental technicism" that subsumes life under the "epistemic, moral, political and ontological scope of the prevailing technicist design-culture".[102] Its static-linear understanding of life does not take into account the dynamics and instabilities of genetic interaction that are pointed out by GM opponents, and that are at least partly taken into account by the EC.

Consequently, "recombinant biopolitics" cannot be depicted as the dominant paradigm of politics for all sites of global power relations. The concept is the result of Dillon and Reid's excellent analysis of contemporary global military-technological discourses, but it cannot be generalised as *the* project of global liberal governance as such.[103] It seems that, in genetics, the "complexity" of the world is still "reduced", while in Dillon and Reid's account of recombinant biopolitics it is already "orchestrated".[104] The effects of Power with regard to biopolitics are both time- and site specific—what becomes an effect of Power in the case of military discourses becomes a mechanism of resistance in the case of GMOs.

Even though a complex-dynamic understanding of life interaction seems to underlie the argument and regulatory practices of the EC, one central assumption of the conventional scientific approach is upheld: the belief that through an inclusion of a wider range of advanced assessment methods and an extension of the territorial and temporal scope of the monitoring and evaluation, the impact of the individual GMO is still controllable and manageable. Paradoxically, the alternative approach of the EC does not lead to the maintaining/sustaining, let alone the opening up, of spaces where there is less regulation (=the "natural environment") but to the confirmation of an expanded and infinite need of control. It is infinite because the effects of the release of GMOs into the environment might be "delayed", possibly long after the "termination of the release", which requires unlimited observation, monitoring and "management".[105] The obvious concern about the long-term effects of GMOs gives a new significance to Foucault's statement that security works on the future, in so far as good planning takes into account "precisely what might happen".[106]

100. *Ibid.*, p. 70.

101. Institute of Science in Society Report, *Death of the Central Dogma* (3 September 2004), available: <http://www.i-sis.org.uk/DCD.php> (accessed 10 June 2009).

102. Wynn, *op. cit.*, p. 75.

103. Michael Dillon and Julian Reid, "Global Liberal Governance: Biopolitics, Security and War", *Millennium*, Vol. 30, No. 1 (2001), pp. 41–65. See also *idem*, *The Liberal Way of War*, *op. cit.*

104. Cf. Dillon and Reid, "Global Liberal Governance", *op. cit.*, p. 51.

105. See, for example, European Communities, *Directive 18/2001/EC*, *op. cit.*, Preamble (20): "Monitoring of potential cumulative long-term effects should be considered as a compulsory part of the monitoring plan." See also *ibid.*, Annex II, Preamble: "delayed effects ... may not be observed during the period of the release of the GMO, but become apparent as a direct or indirect effect either at a later stage or after the termination of the release".

106. Foucault, *Security, Territory, Population*, *op. cit.*, p. 20.

These paradoxical conclusions need to be related to my analysis of the particular neoliberal policy of the EC with regard to GMOs, because the same complex-dynamic understanding that underlies its approach to science provides the basis for so-called "evolutionary economics" that draws on the work of Friedrich von Hayek. What needs to be taken into account is the particular problem this new scientific and economic understanding poses for governmental bodies, which need to redefine their role in the management of biological and economic life if both are understood to be developing in non-linear, unpredictable ways.

In one of the rare studies of the consequences of "evolutionary economics" for governance, Kurt Dopfer's edited volume *Economics, Evolution and the State: The Governance of Complexity*, it is precisely this tension that is dealt with in various ways. One of the contributors, Christian von Weizsäcker, argues that evolutionary economics can only seriously challenge the predominance of linear, function-based neoclassical economics if it manages to develop an "evolutionary welfare economics". This requires connecting "positive statements" to "normative statements", which means evaluating existing policies and drafting policy recommendations based on positive assumptions about economic processes.[107] Following Wolfgang Kerber's analysis in the same volume, the difficulty of conceptualising positive policy has to do with the "openness of economic processes and the resulting Hayekian knowledge problem" on which evolutionary reasoning is based.[108] As Weizsäcker shows, crucial for this is the possibility of the measurement of value, for which in neoclassical economics the standard method is monetarisation.[109]

Even though all of the contributors to Dopfer's volume attempt to solve this dilemma, they are not able to fundamentally overcome it. Attempts to combine governance and evolutionary economics are mostly in need of a "pragmatic compromise", as Kerber stresses in his three principles for evolutionary economic research:

> 1. Evolutionary economics should be pragmatic in a methodological sense, i.e. that in an ever-changing world *economic policy can and has to be made* despite the impossibility of eliminating all uncertainties in regard to its effects. 2. Evolutionary economics should not restrict itself to theoretical and basic research but should also do research about the application of evolutionary reasonings to the *solving of real-life problems including participation in policy discussions*. 3. For applying evolutionary arguments to policy questions we have to find a *pragmatic way* to *combine evolutionary with neoclassical arguments*, which to a certain degree will *remain indispensable* for many real-world problems.[110]

Interestingly, the EC deals with the dilemma of combining a complex-dynamic scientific approach with governmental policy by using disciplinary techniques. In its regulation of the release of GMOs into the environment, it follows a

107. C. Christian von Weizsäcker, "Is the Notion of Progress Compatible with an Evolutionary View of the Economy?", in Dopfer (ed.), *op. cit.*, pp. 43–44.

108. Wolfgang Kerber, "Applying Evolutionary Economics to Public Policy—The Example of Competitive Federalism in the EU", in Dopfer (ed.), *op. cit.*, p. 296.

109. von Weizsaecker, *op. cit.*, p. 44.

110. Kerber, *op. cit.*, p. 296; my emphasis.

procedure that resembles what Foucault calls the third stage of discipline, which is the adding up and capitalisation of time. The EC argues, for example, that "the introduction of GMOs in the environment should be carried out according to the step by step principle". This means that "the containment of GMOs is reduced and the scale of release increased *gradually, step by step*, but only if *evaluation* of the earlier steps in terms of protection of human health and the environment indicates that the *next step* can be taken".[111] Similarly, according to Foucault, the adding up and capitalisation of time implies the "division of duration" into "successive or parallel segments", the analytical organisation of the segments, and their finalising with examination.[112] Moreover, it is striking how much the required system of classification, the description of processes and interactions, and the methods of monitoring and control are like Foucault's description of the disciplines in the "plague town"; particularly with regard to "emergency response plans" in the case of an "unexpected spread" of GMOs.[113]

The EC's information and labelling policy, which is part of a Regulation that was designed to regulate the import of GM products, also highlight the need for disciplinary techniques such as characterisation and classification.[114] Apparently, part of the EC's approach towards the governance of GMOs is based on the assumption that with the help of disciplinary techniques, complex unstable natural processes and interactions are still controllable and manageable. This is similar to what Nikolas Rose refers to as the "molar" reaction of the World Health Organization to the 2003 outbreak of SARS, which took the form of "quarantine, travel restrictions, and the policing of space that has been familiar since the medicine of the plague".[115]

Instead of following the techniques of the "norm" and calculating the probability of a certain "contamination" of the environment through GMOs, and thereby the *relative* danger,[116] the EC pursues a containment strategy that is concerned about the *absolute* danger that GMOs might pose. This is part of the so-called precautionary approach that I have already referred to. The use of techniques of discipline does exist in a complex relationship with the EC's neoliberal understanding of providing "choice" for the consumer. In Deleuzian terms, the EC rigorously "reterritorialises" GMOs with the help of disciplinary techniques. The complexity of life is reduced through "strategies of compartmentalization and spatial localization" and thereby transformed into an object of governmental action.[117] Through these moves, GMOs become governable and thereby enable a reterritorialisation of political rule itself.

All components of the Foucauldian "triangle"—(neo-)liberal governmentality, sovereignty, and discipline—are now brought into the picture. The analysis has elucidated their complex interrelationship, mutual constitution and internal

111. European Communities, *Directive 18/2001/EC, op. cit.*, Preamble (24); my emphasis.

112. Michel Foucault, *Discipline and Punish: The Birth of the Prison* (London: Penguin, 1979), pp. 157–158.

113. European Communities, *Directive 18/2001/EC, op. cit.*, Annex III A, C (V). Cf. Lezaun, *op. cit.*, who comes to the same conclusion.

114. European Communities, *Regulation (EC) 1830/2003*. Cf. Foucault, *Discipline and Punish, op. cit.*, p. 203.

115. Nikolas Rose, *The Politics of Life Itself: Biomedicine, Power, and Subjectivity in the Twenty-first Century* (Princeton and Oxford: Princeton University Press, 2007), p. 13.

116. Cf. Ewald, *op. cit.*, p. 142.

117. Lezaun, *op. cit.*, p. 514.

tensions in a specific case, and has thereby enabled insights into certain functions and procedures of contemporary global governance.

Conclusion

In *Security, Territory, Population* Foucault stresses that in order to avoid reducing the exercise of power to a single logic, it is necessary to analytically perform a "triple displacement ... to the outside": First, "moving outside the institution, moving off-center in relation to the problematic of the institution", second, substituting "the external point of view of strategies and tactics for the internal point of view of the function", and third, "grasping the movement by which a field of truth with objects of knowledge was constituted through these mobile technologies".[118] If this shift to the outside is exercised with regard to global power relations, the trap of depicting it as a homogeneous project is avoided. This is exactly what I have attempted to do in this contribution with regard to a particular case of what is commonly perceived as "global governance".

With his approach of decentring, Foucault enables students in IR to recognise the variety of tactics, strategies, field of truths and rationalisations that can be found in any particular regime of power/knowledge. His insistence on the significance of heterogeneity makes it possible for us to regard all constellations of power, whether at the international, national, or any other level of analysis, as uneven and irreducible to particular concepts. It is exactly this unevenness, which emerges at all micro- and macro levels, that reveals the inadequacy of general theories that smooth out complexities, tensions, mobilities and instabilities in social interactions. Moreover, it opens up the field of politics to contestation and new possibilities; without losing sight of solid Power effects.

118. Foucault, *Security, Territory, Population*, *op. cit.*, pp. 116–118.

"... we are being left to burn because we do not count"*: Biopolitics, Abandonment, and Resistance

ANNA SELMECZI**

Starting from the puzzle posed by the ultimate aim of modern governmental rationality to nurture the population and its tendencies to exclude large parts of the same population from the spectrum of its care, this article argues that abandonment is always already inscribed into this rationality. In contradiction to Agamben, abandonment here is not attributed to the sovereign exception but is traced back to modern processes transforming the political—as problematised by Hannah Arendt and Michel Foucault. Complementing their observations with the empirical and the anti-political implications of "the count" based on Ian Hacking's and Jacques Rancière's thought, first a conceptual framework for understanding biopolitical abandonment is outlined, then the materialisation of abandonment is assessed. Arriving finally at the possibility of thinking resistance to the power that disallows life through conceiving of politics as disruption, the final section discusses the South African shack-dwellers' struggle that, on occasions, is able to disturb the dynamics of abandonment and so potentially furthers the conceptualisation of resistance to biopolitics.

> ... the people is those who, refusing to be the population, disrupt the system.[1]

Introduction

Michel Foucault once said that one of the biggest antinomies of modern political reason was the coexistence of "large destructive structures and institutions oriented toward the care of individual life".[2] Indeed, if modern governmental

* Abahlali baseMjondolo, "Fire Devastates the Kennedy Road Settlement—At Least One Hundred Homes are Destroyed", Press Release (16 June 2009).

** The author would like to express her gratitude to her supervisor, Michael Merlingen and the members of the Abahlali baseMjondolo for all their support. She would also like to thank Nicholas J. Kiersey and Jason R. Weidner for organising the ISA 2009 panel on which an earlier version of this paper was presented, and Nicholas Kiersey in particular for his efforts and support throughout the editing of this issue. Finally, the author is grateful to Alexander Astrov, Xymena Kurowska, Adam Mestyan and Erzsebet Strausz for their comments on earlier drafts.

1. Michel Foucault, *Security, Territory, Population: Lectures at the Collège de France, 1977–78* (Houndmills and New York: Palgrave Macmillan, 2007), p. 66.

2. *Idem*, "The Political Technology of Individuals", in Luther H. Martin, Huck Gutman and Patrick H. Hutton (eds.), *Technologies of the Self* (Amherst: University of Massachusetts Press), p. 147.

rationality includes a strong element of pastoral care and centres on the well-being of each and all, then what should we make of the extreme marginalisation of millions inherent in the dynamics of global neoliberal governance? How is it possible that a rationality that defines its ultimate aim as the nurturing of the population nevertheless accommodates the abandonment of huge masses of people? The answer seems to lie in understanding the ways in which abandonment is always already inscribed into it. This paper aims to provide such an understanding and, by way of pointing to instances where the inscribed abandonment is challenged, attempts to offer a possible (although as yet rudimentary) conceptualisation of resistance against biopolitical governance. This problematic seems all the more important in the context of recent contributions to a debate within the discipline of International Relations (IR) about the potentials of Foucault's thought and the biopolitics literature drawing on it. Whereas this literature certainly has its limits—the failure to provide a conception of resistance so far seems to be one of these—recourse to a more traditional, state-centred, or territorial theorisation of the political seems even less likely to address present-day problems of marginalisation and community. Therefore, in what follows, starting from Hannah Arendt's and Foucault's observations on the modern transformation of the political, then focusing on the empirical (Ian Hacking) and the (anti-)political (Jacques Rancière) implications of "the count", in the first two sections I construct a conceptual framework for understanding biopolitical abandonment. Having drawn the conceptual outlines, based on Rancière's aesthetic approach to politics, the materialisation of abandonment is assessed. Finally, arriving at the possibility of thinking resistance to the power that disallows life through Rancière's concept of politics as disruption, the final section discusses the South African shack-dwellers' struggle that, on occasions, appears to be able to disturb the dynamics of abandonment.

One of the existing attempts to understand how abandonment is inherent in modern rationalities of rule is Giorgio Agamben's, who takes up Foucault's concept of biopolitics—according to which the governmental concern for the well-being of the people implies that the human as a biological being enters the realm of the political—but argues that politics and biological life had always been tied together. For Agamben, modernity only sheds light on and reaffirms the fact that "the inclusion of bare life into the political realm constitutes the original—if concealed—nucleus of sovereign power".[3] Therefore, he claims, it is crucial to study sovereign and biopolitical models of power in parallel, because it is in the very intersection of these two—in the ambiguous zone of the sovereign exception—that we can locate bare life: the life that can be taken without committing homicide and to which we are all reduced contemporarily.

That Agamben's theorisation of modernity through the notion of exception and bare life resonates with perceptions of our present is reflected (among others) in poststructuralist IR literature which, in recent years, eagerly took up these conceptions. Accounts that took as their analytical point of departure Agamben's claim that the (concentration) camp is the paradigm of modernity were able to show how bare life is produced through the obscured interplay of sovereign power and biopolitics on sites such as the refugee camp or the detainment

3. Giorgio Agamben, *Homo Sacer: Sovereign Power and Bare Life* (Stanford, CA: Stanford University Press, 1998), p. 6.

camp.[4] Doubtlessly, *homo sacer* is an expressive concept when interrogating the operation of power in such loci. However, assuming that the operation of biopower inevitably leads to the camp and bare life produced in it can also limit our understanding of the contemporary because this assumption, through forging a connection between the sovereign exception and biopower, generalises a distribution of power that Foucault referred to as the state of domination.[5] As opposed to relations of power that, in Foucault's conceptualisation,[6] can always be reversed and that are always conditioned upon the freedom of its parties to this reversal, states of domination hardly allow for resistance.[7] Accordingly, accounts of resistance in refugee or detainment camps disclose forceful but limited practices of resistance taking shape in individual performances such as lip-sewing and refugee poetry.[8]

Yet biopower is not at all bound to operate in relations of domination. On the contrary, a vast part of the literature based on Foucault's conceptualisation of modern power and, in particular, his conception of biopolitics, analyses how biopolitical technologies of government are deployed through the freedom of the governed.[9] Observing this aspect does not necessarily mean disregarding the dark side of biopolitics. Indeed, Foucault famously defined modern power as "the power to *foster* life or *disallow* it to the point of death".[10] Therefore, in an attempt to understand in what ways abandonment is inscribed in modern governmental rationality and consequently to be able to conceptualise resistance to contemporary biopolitical abandonment without equating biopolitical abandonment with the sovereign exception (the sovereign ban),[11] I suggest using a different focal

4. See, for example, Jenny Edkins, "Sovereign Power, Zones of Indistinction, and the Camp", *Alternatives: Social Transformation & Humane Governance*, Vol. 25, No. 1 (2000), pp. 3–27; Jenny Edkins and Veronique Pin-Fat, "Through the Wire: Relations of Power and Relations of Violence", *Millennium: Journal of International Studies*, Vol. 34, No. 1 (2005), pp. 1–24.

5. Michel Foucault, "The Ethics of the Concern for the Self as a Practice of Freedom", in Paul Rabinow (ed.), *Essential Works of Foucault*, Vol. I: *Ethics* (London: Penguin, 2000), pp. 281–301.

6. Michel Foucault, "Subject and Power", *Critical Inquiry*, Vol. 8, No. 4 (1982), pp. 777–795.

7. On Foucault's distinction between relationships of power, states of domination, and government (an intermediary category understood broadly as the conduct of conduct) see Barry Hindess, *Discourses of Power: From Hobbes to Foucault* (Oxford: Blackwell, 1996), pp. 96–136. It is based on the Foucauldian conception of power as "action upon the actions of others"—or, as Hindess formulates it, as "an ubiquitous feature of human interaction"—that Laura Zanotti criticises Agambenian interpretations of liberalism. In line with what is stated above, Zanotti argues that the government of disorderly states does not "produce totalizing effects of domination". Instead, through conducting the conduct of states to be disciplined, normalisation inscribes spaces of resistance that allow for diverting and hijacking its original agendas, as in the case of the international attempts to secure order in Croatia. Cf. Laura Zanotti, "Normalizing Democracy and Human Rights: Discipline, Resistance, and Carceralization in Croatia's Euro-Atlantic Integration", *Journal of International Relations and Development*, Vol. 11, No. 3 (2008), pp. 222–250.

8. See Edkins and Pin-Fat, *op. cit.*; Prem Kumar Rajaram, "Disruptive Writing and a Critique of Territoriality", *Review of International Studies*, Vol. 30, No. 2 (2004), pp. 201–228.

9. See, for example, Nikolas Rose, *Powers of Freedom: Reframing Political Thought* (Cambridge: Cambridge University Press, 1999); Andrew Barry, Thomas Osborne and Nikolas Rose, *Foucault and Political Reason: Liberalism, Neo-liberalism, and Rationalities of Government* (Chicago, IL: University of Chicago Press, 1996).

10. Michel Foucault, *The History of Sexuality, Volume 1: The Will to Knowledge* (London: Penguin, 1998), p. 138; emphasis in original.

11. While I do not engage here in a detailed discussion of Agamben's notion of abandonment as the sovereign exception, it is not my intention to dismiss it completely. As mentioned below in relation to the notion of superfluity, and as the above references to analyses of power and resistance in

lens than that of Agamben (that is the "bare life"), although my point of departure is, similarly, Arendt's and Foucault's observations on the biologisation of the political.[12]

The Naturalisation of the Political

Choosing to proceed through this alternative perspective implies that the present discussion problematises characteristically modern phenomena: both biopolitical abandonment and the governmental rationality into which it is inscribed emerged with modernity and neither of them is (directly) tied to ancient sovereign models of power. Thus, the primary question has to address the shift in the forms of power occurring with modernity. What is it in modernity that forces sovereign power to retreat and makes way to a mode of power that either fosters life or abandons it? As hinted at above, enabling this form of power is an assemblage which, based on Arendt's and Foucault's parallel observations, André Duarte refers to as the "naturalization of the political".[13] This process, the ultimate outcome of which is that the political is refocused on the biological existence of humanity,[14] comprises a series of naturalisations, that is, the serial reconceptualisation of certain phenomena as driven by the laws of nature.

If we want to reconstruct the sequence of this series (bearing in mind that its elements are in constant interaction), it seems appropriate to start with the emergence of economy, that is, when economy had begun to be thought about as an autonomous field of intervention and as a fundamental organising principle and was complemented by its correlative forms of knowledge, political economy in particular. For Arendt, the emergence of economy coincides with

contemporary Camps show, this concept indeed has relevance in certain situations. Nevertheless, its relevance cannot be extended to all manifestations of biopolitical abandonment, for these, I believe, are more often inscribed not into states of domination but into governmental rationalities and practices characteristic of biopolitical models of power. This is precisely what enables and at once necessitates thinking resistance to their inscription. As Didier Bigo argues, Agamben criticises Foucault's very conception of the indivisibility of power and resistance: "For him, and contrary to Foucault, the polarization between power and bare life is possible and in fact drives all the contemporary practices of power, including those of liberal states and democracies." The conception of this polarisation is made possible by Agamben's reduction of Jean-Luc Nancy's notion of the ban. This reduction, according to Bigo, implies that "by exaggerating the capacity of the actors speaking in the name of the sovereign and by essentialising sovereignty through a conception that plays against (yet with) the rule of law [. . .] Agamben ignores the resistance of the weak and their capacities to continue to be humane and to subvert the illusory dream of total control". Didier Bigo, "Detention of Foreigners, States of Exception, and the Social Practices of the Control of the Banopticon", in Prem Kumar Rajaram and Carl Grundy-Warr (eds.), *Borderscapes: Hidden Geographies and Politics at Territory's Edge* (Minneapolis and London: University of Minnesota Press, 2007), pp. 3–33. For a criticism of Agamben's "political nihilism" that entails dismissing "all political options in our societies" see further Ernesto Laclau, "Bare Life or Social Indeterminacy?", in Matthew Calarco and Steven DeCaroli (eds.), *Giorgio Agamben: Sovereignty & Life* (Stanford: Stanford University Press), pp. 11–22.

12. See Agamben, *op. cit.*, p. 120.

13. André Duarte, "Biopolitics and the Dissemination of Violence: The Arendtian Critique of the Present", HannahArendt.net, available: <http://hannaharendt.net/research/biopolitics.html> (accessed 21 June 2009).

14. See the most quoted "definition" of biopolitics: "For millennia, man remained what he was for Aristotle: a living animal with the additional capacity for a political existence; modern man is an animal whose politics places his existence as a living being into question" (Foucault, *The History of Sexuality, op. cit.*, p. 143).

the construction of society, and so contributes to the blurring of the line of division between the private and the public realms.[15] Certainly, as will be shown in this paper too, the conception of anciently distinct private and political realms is problematic. Nevertheless, as Patricia Owens writes, Arendt "did not believe that the public and private sphere were entirely unrelated"; "she argued that there are distinct principles and motives for political action".[16] While these claims still need to be assessed from the perspective defined by the conception of the political that the discussion below draws on, Arendt's following observation is still important: with economy being applied to the "super-human family" of a modern state and the outlines of a "collective housekeeping" being drawn, biological necessities of the human are channelled into the political realm. At the core of this development is the assumption of a "'collective' concern" that includes the concern for the well-being of society as a whole.[17] Indeed, in citing Gundar Myrdal, Arendt states: "economics can be a science only if one assumes that one interest pervades society as a whole".[18] For Arendt, this "communistic fiction" of liberal utilitarianism implies nothing less than the subjection of the human potential for action to a universal behavioural pattern. This pattern, in turn, lends itself to forms of scientific knowledge that operate by and produce predictions on the large scale, so rendering individual action invisible and outstanding deeds deviant.

Claiming that the general framework of biopolitics is liberalism (understood as an art of government), Foucault reconstructs a very similar process of transformation.[19] Hence, it is the *économiste* critique of mercantilism to which he traces back the emergence of biopolitics. In straight opposition to mercantilist restrictions intended to avoid the harmful effects of the market, classical economists of the 18th century viewed the market as directed by natural forces that, when unobstructed, strive for harmony. Therefore, instead of intervening in their flows, proper government must let "things follow their course".[20] Through such notions as the "true price" that is formulated by natural economic dynamics, the market is now reconstructed as a site of *veridiction* and the sovereign is thought of as incapable of fully grasping the truth of the market. Consequently, "the principle of the self-limitation of governmental reason" appears.[21]

Crucially, this self-limitation based on the naturalisation of the market brings about the naturalisation of the subject of government. When individuals come to be seen as linked to the natural economic processes through their "longevity, health, and ways of conducting themselves", a new target of government is formed: the population.[22] Once the attention of the sovereign is refocused from the individual—as the subject of rights but also as the subject/object of disciplinary power—to the populational level, rationality and practices of power change radically. The sovereign gaze—that now becomes governmental—will no longer

15. Hannah Arendt, *The Human Condition* (Chicago, IL: University of Chicago Press, 1958), pp. 28–29.

16. Patricia Owens, "Hannah Arendt", in Jenny Edkins and Nick Vaughn-Williams (eds.), *Critical Theorists in International Relations* (London and New York: Routledge, 2009), p. 37.

17. Arendt, *op. cit.*, p. 33.

18. *Ibid.*, p. 44.

19. Michel Foucault, *The Birth of Biopolitics: Lectures at the Collège de France, 1978–1979* (Houndmills and New York: Palgrave Macmillan, 2007), p. 22.

20. *Idem, Security, Territory, Population, op. cit.*, p. 48.

21. *Idem, The Birth of Biopolitics, op. cit.*, p. 20.

22. *Ibid.*, p. 22.

focus on the actions and behaviours of individuals, but on the natural processes of the population into which individuals are massified by the forms of knowledge attached to political economy. Here again, the intention to regulate the aggregate processes of the population implies the assumption of interest being the sole motivating force: a universal, through which this conglomerate governmental subject/object is accessible and can be governed towards its well-being. "[T]he population taken as a whole has one and only one mainspring of action. This is desire."[23] Beyond designating "the entry of a 'nature' into the field of techniques of power", the collective interest of the population, which adds up from the interests of its benefit-seeking elements, becomes the ultimate end of government.[24]

It is exactly here that Foucault's aphoristic claim about the major transformation of the forms of power is situated. At the core of this claim is the shift in the main problem of those who govern: with the market reconstructed as a site of truth-telling and with the desires of the population reconstructed as the forces defining the collective interest (which, in turn, is conceptualised as the end of government) sovereign power inevitably retreats. Therefore, the central problem of government will no longer be how and on what legal bases the sovereign can say no to the subject of right under its rule, but how to say yes to desire. When the sovereign no longer has direct hold on its subjects (including both their belongings and their lives) and when the aim of governmental power becomes the regulation of economically relevant processes through the population and for the population, then indeed it seems that "the ancient right to *take* life or *let* live was replaced by a power to *foster* life or *disallow* it to the point of death".[25] Thus, the power to foster life implies the proper stimulation or encouragement of interests present in the massified subject/object of the population.

But what does it mean to disallow life to the point of death? What, in fact, could be thought of as "negative biopolitics"?[26] In order to answer this question we have to revisit the interrelated implications of the naturalising processes discussed above. On the one hand, and according to Arendt, with the subsumption of human actions to a behavioural pattern defined by the collective concern, the plurality and contingency of individuality is eliminated from the construction of the (neither private nor public) social sphere:

> It is decisive that society, on all its levels, excludes the possibility of action, which formerly was excluded from the household. Instead, society expects from each of its members a certain kind of behavior, imposing innumerable and various rules, all of which tend to "normalize" its members, to make them behave, to exclude spontaneous action or outstanding achievement.[27]

23. Foucault, *Security, Territory, Population, op. cit.*, p. 72.

24. *Ibid.*, p. 75.

25. Foucault, *History of Sexuality, op. cit.*, p. 138; original emphasis. It is exactly this shift that cannot be captured when accepting Agamben's claim on the ancient bond between sovereign and biopolitical models of power.

26. "Perhaps if Foucault could have seen the way African 'demography' is 'regulated' by the AIDS epidemic (and a number of other epidemics, all monitored by a 'World Health Organization'), he might have ventured to speak of 'negative bio-politics'" (Étienne Balibar, *Politics and the Other Scene* (London and New York: Verso, 2002), p. 38).

27. Arendt, *op. cit.*, p. 40.

In Foucauldian terms and from the perspective of the art of government, the elimination of the individual occurs (also) through the discussed rescaling of the target of government. It is now the processes at the populational level which are pertinent, and not the actions or behaviours occurring at the level of individuals or the multiplicity of individuals. This implies that the sovereign attention that still persisted in the omnipresent disciplinary gaze focusing on the individual body is, at least at the primary level of government, replaced by a generalising gaze of a massifying power. Betterment is now supposed to take place at the level of "life in general".[28]

Betterment of life at this level of generality, in turn, is unimaginable without the forms of knowledge that are meant to grasp it in its totality. Biopower is, after all, that which brought "life and its mechanisms into the realm of explicit calculation and made knowledge-power an agent of transformation of human life".[29] Economics, demography, epidemiology and, most importantly perhaps, statistics, provided the indivisible epistemic element for these two eliminatory aspects of the biopolitics of "man-as-species".[30]

> The laws of statistics are valid only where large numbers or long periods are involved, and acts or events can statistically appear only as deviations or fluctuations [. . .] The application of the law of large numbers and long periods to politics or history signifies nothing less than the wilful obliteration of their very subject matter, and it is a hopeless enterprise to search for meaning in politics or significance in history when everything that is not everyday behavior or automatic trends has been ruled out as immaterial.[31]

Foucault's more nuanced meditation on the differences between the relationship of the regulatory biopolitics of the man-as-species and the disciplinary biopolitics of the man-as-body to the "norm" and the "normal" leads to similar conclusions. Whereas disciplinary techniques are based on the differentiation of the normal individual from the abnormal individual according to a set norm, regulatory techniques deduct the normal distribution of cases from statistical trends, and if the actual distribution diverges from the normal then it is this composite line of divergence that has to be acted upon. Indeed, as Ian Hacking argues, when society became statistical, the notion of "normal people" gained shape—with enormous consequences:

> People are normal if they conform to the central tendency of [social] laws, while those at the extremes are pathological. Few of us fancy being pathological, so "most of us" try to make ourselves normal, which in turn affects what is normal.[32]

It is thus almost impossible to underestimate the significance of the emergence of statistical knowledge.[33] What the will to statistical knowledge brought along was

28. Michel Foucault, *"Society Must be Defended": Lectures at the Collège de France, 1978–1979* (Houndmills and New York: Palgrave Macmillan, 2007), p. 241.
29. *Idem*, *History of Sexuality*, *op. cit.*, p. 143.
30. *Idem*, *"Society Must be Defended"*, *op. cit.*, p. 242.
31. Arendt, *op. cit.*, pp. 42–43.
32. Ian Hacking, *The Taming of Chance* (Cambridge: Cambridge University Press, 1990), p. 2.
33. Or, to borrow Hacking's term, "the avalanche of printed numbers" (*ibid.*).

nothing less than the absolutisation or, put better, the literalisation of the count. What does this slightly farfetched phrase refer to? Clearly, if the primary aim of the fetishistic wave of data collection in the first half of the 19th century tackled by Hacking's analysis was acquiring knowledge of and intervening in phenomena related to the life of the population, its primary effect was categorisation. "Enumeration demands *kinds* of things or people to count. Counting is hungry for categories."[34] Once categories are defined, not one case of death, disease, or profession will escape clustering; and this is so even if categories themselves change.[35] "[B]ureaucrats [...] designed easily countable classifications, into which everybody had to fall—and thenceforth did".[36]

This is where "literalisation of the count" falls into place and Jacques Rancière's work appears to be worthy of much more than an intertextual reference. In his conceptualisation of the paradoxical relationship between politics and philosophy, political philosophy features as always aiming to impede the occurrence of the political by accounting for all parts of the community and distributing forms of participation accordingly. That is, according to Rancière, political philosophy—and thus the common good—is conditioned upon tying political idealities to the appropriate parts of the community; upon "a count whose complexities may mask a fundamental miscount [...] the very wrong that is the stuff of politics".[37] As we will see below, politics is always the processing of this fundamental miscount: it is always the confrontation of the supposedly total count of the police order with a part that is unaccounted for. What is to be considered at this point of our discussion is, however, this: what happens to the miscount when modern social science appears? In Rancière's interpretation, when social science of the 19th century emerges and becomes the form of existence political philosophy has taken ever since, politics is eliminated. When "exhaustive counting mechanisms" are introduced, the sum of its parts will finally be equal to the people. "Such a people, present in the form of its statistical reduction, is a people transformed into an object of knowledge and prediction that sends appearance and its polemics packing."[38] The count thus reaches perfection when it becomes literal.

Taking a look at the earliest phase of this process through the example of the classification of occupations will be illustrative. Hacking notes that in the beginning of the 19th century, the classification of people in terms of occupation changed radically: categories were formulated with the aim of facilitating the all-encompassing count. While the industrial revolution in itself could justify a drive for the enumeration of proliferating professions, the timing of "statistical enthusiasm" between two revolutionary years (those of 1832 and 1848) suggests that the desire to count everything was conceived of as a means of containing

34. Ian Hacking, "Biopower and the Avalanche of Printed Numbers", *Humanities in Society*, Vol. 5 (1982), p. 280; original emphasis.

35. "[N]ational and provincial censuses amazingly show that the categories into which people fall change every ten years. Social change creates new categories of people, but counting is no mere report of developments. It elaborately, often philanthropically creates new ways for people to be" (Ian Hacking, "Making up Individuals", in Thomas C. Heller, Morton Sosna and David E. Welbery (eds.), *Reconstructing Individualism: Autonomy, Individuality, and the Self in Western Thought* (Stanford: Stanford University Press, 1986), p. 223.

36. Hacking, "Biopower and the Avalanche of Printed Numbers", *op. cit.*, p. 280.

37. Jacques Rancière, *Dis-agreement: Politics and Philosophy* (Minneapolis: University of Minnesota Press, 1999), p. 6.

38. *Ibid.*, p. 105.

insurgency: "Find out more about your citizens, cried the conservative enthusiasts, and you will ameliorate their conditions, diminish their restlessness, and strengthen their character."[39] The echo of Foucault's notion of biopolitics in this paraphrase is not, of course, arbitrary. Neither can thus be the episode cited by Rancière to illustrate a speech scene within which political subjectification disrupts the police order—as the textual locus of disruption is a concrete classification: that of recognised professions. When in 1832 revolutionary Auguste Blanqui declared to the jury that his profession is "proletarian" and redefined "profession" as "a profession of faith, a declaration of membership of a collective", the count of the police order had been disturbed by the political subjectification of a part that had been unaccounted for.[40] The proletariat was not equivalent to a social group, neither did it become a profession once the judge, accepting Blanqui's redefinition of the term, ordered adding "proletarian" to the list of professions. Understanding this speech event as a disruption of the police order seems to be justified when considering what, according to Rancière, the role of "occupation" is within ordering: "The distribution of the sensible reveals who can have a share in what is common to the community based on what they do and on the time and space in which this activity is performed."[41] With political philosophy giving way to a social science that thus vastly outshone the wildest dreams of statistical enthusiasts, miscounts based on which such scenes can be staged were largely effaced: politics had been eliminated.

Abandoning the Superfluous

It would be rather odd to claim that the above interpretation of Arendt, Foucault, and Rancière add up to a perfect prism, looking through which will provide a full view of the way abandonment is inscribed into modern governmental rationality. However, a parallel reading of their conceptions on how, in interaction with the emergence of the social, political philosophy and governmental rationality has been transformed and what this implies, productively enlarges our perspective on the workings of this inscription.[42] If we had to condense the composite picture we thus gain, its key element would certainly be the emergence of the aggregate subject/object of government: the "super-human family" or the population. In Foucault's genealogy we have seen that the appearance of this massified domain resulted in the rescaling of governmental rationality and practices: while neither sovereign nor disciplinary modes of power disappear, in the primary focus of governmental concern are now natural processes that are pertinent only at the level of the population; trivially implying that processes below this level are of no pertinence. Whereas Foucault explicitly chooses to trace the emergence of biopower "not at the level of political theory, but rather

39. Hacking, "Biopower and the Avalanche of Printed Numbers", *op. cit.*, p. 281.

40. Rancière, *op. cit.*, p. 38.

41. *Idem, The Politics of Aesthetics* (New York: Continuum, 2004), p. 12.

42. Thus, I do not wish to suggest that there is a continuum of thought around the notions of abandonment or superfluity in the work of the authors discussed; in fact, at points, there are significant tensions between them. While some of these are mentioned below, the discussion of discontinuities is not the object of this article. My aim here is rather to place the referred authors' concepts into interaction, so that they illuminate each other and thus shed some light on aspects of the problematic of abandonment that has remained obscure in current discussions.

at the level of the mechanisms, techniques, and technologies of power",[43] through nearing his analysis to Arendt's and Rancière's thought we also get a view from the perspective of the former level.[44] From this angle, as was hinted at above, we encounter the incommensurability of the political and the statistical reduction called population.

The super-human family is incapable of spontaneous action: its behaviour is determined by social laws.[45] Identical to the sum of its parts, the population is incapable of creating scenes of appearance: it is walled up into a "homogeneous regime of the visible".[46] We must not overlook the implications of this homogeneity as, on the one hand, it represents the main feature of the post-democratic order, that is, of the order that emerges after political philosophy takes the form of social science and any acting out of *dissensus* is ruled out by the consensus of public opinion. When everyone "is included in advance, every individual is the nucleus and image of a community of opinions that are equal to parties", there is no representable barrier.[47] Unlike previous ages, which unabashedly divided those worthy of political life from the dumb rest, the consensus-discourse that wages war on "exclusion" impedes the polemical construction thereof.[48]

Taking a look at the level of governmental rationality and practices again, we find an early parallel to the consensual overwriting of representable barriers in the form of the modern readjustment of the divisions that determine proper government. As opposed to the primary problem of the age of *raison d'état*, once the market had been reconstructed as a site of *veridiction*, the question facing the sovereign will not be whether his rule is legitimate or it violates the rights of its subjects. The internal limitation against which the formulating liberal governmental reason keeps running up establishes a new division:

> [T]his governmental reason will not divide subjects between an absolutely reserved dimension of freedom and another dimension of submission which is either consented to or is imposed. In fact, the division is not made within individuals, men, or subjects, but in the very domain of governmental practice, or rather within governmental practice itself, between the operations that can be carried out and those that cannot, between what to do and the means to use on the one hand, and what not to do on the other.[49]

Consequently, the decision between what must be done and what should be left to its own dynamics is conceived of not as the object of sovereign decision but as directed by the nature of the objects of governance, the truth of which, as mentioned above, the sovereign would be incapable of comprehending and

43. Foucault, *"Society Must be Defended"*, *op. cit.*, p. 241.

44. On the Arendtian aspect of Rancière's thought see Jean-Philippe Deranty, "Rancière's Political Ontology", *Theory and Event*, Vol. 4, No. 4 (2003).

45. Even if these laws are probabilistic and contain an element of contingency, so constituting the crux of security apparatuses. On this aspect of biopolitical governance see Michael Dillon, "Governing through Contingency: The Security of Biopolitical Governance", *Political Geography*, Vol. 21, No. 3 (2007), pp. 41–47.

46. Rancière, *Dis-agreement*, *op. cit.*, p. 104.

47. *Ibid.*, p. 116.

48. For a bizarrely nostalgic description of the contrasting rationalisation of exclusion in the past see Rancière, *ibid.*

49. Foucault, *The Birth of Biopolitics*, *op. cit.*, p. 11.

controlling. Frugal government of the "modest state" leaves limited choices to both governed and governor. In our post-democratic present, which, in many senses, brought along the perfection of the discussed features of liberal governmental rationality, the assumption of powerlessness becomes ever more crucial:

> The legitimacy of state power is [. . .] reinforced by the very affirmation of its impotence, of its lack of choice faced with the world-wide necessity it is dominated by. The theme of the common will is replaced by that of the lack of personal will, of capacity for autonomous action that is anything more than just management of necessity.[50]

Ultimately, if the above redrawn trajectory is tenable, the interrelated emergence of the economic and social realms as autonomous fields of intervention—through the naturalisation of the processes attributed to them—leads to the elimination of contingency from both the subject/object and the act of governance. Thus, to put it in Arendtian terms, human being becomes superfluous.[51] Hence, by way of this term, we arrived at the phenomenon that can be understood as the abandonment inscribed in modern governmental rationality, that is, biopolitical abandonment. But how exactly are biopolitical abandonment and superfluity related? Preserving our twofold perspective and first looking at the realm of governmental rationality, here again we are led to the pertinence of the population and the non-pertinence of anything below its aggregate level. Considering Foucault's discussion of the event of scarcity, however, also clarifies why non-pertinence can at once be understood as superfluity. When scarcity, conceived of as a scourge that affects both the individual and the whole population and must therefore be avoided, is replaced by a conception of scarcity as a chimera that exists at the level of individuals or particular multiplicities of individuals, but does not affect the population as the aggregate object of government, letting people die becomes integrated into governmental rationality. Securing against the emergence of scarcity at the pertinent level of the population means allowing it to develop on particular sites and affecting particular individuals or groups on the non-pertinent level. "The scarcity-scourge disappears, but the scarcity that causes the death of individuals not only does not disappear, it must not disappear."[52] We could thus say that modern governmental rationality that aims at the well-being of the population entails a certain "perspectival superfluity": the particular sites and people on which natural processes are

50. Rancière, *Dis-agreement, op. cit.*, p. 113.

51. For Arendt, superfluity features primarily as the aim of totalitarian regimes, for which, as a result of their aim to speed up the progress of the (human) race towards its historical fate, the human potential for spontaneous action is unnecessary, and so—this potential being what makes it what it is—human itself becomes superfluous. In her view, this aim had only been achieved in the concentration camps where, being reduced to mere corpses, human beings were indeed lacking the capacity for action. (This sense of superfluity can be read as Agamben's *homo sacer.*) Beyond this notion, however, Arendt uses the term in another, more literal sense: referring to stateless people and the millions of unemployed who were excluded from the protected sphere of their nation-states because they were, for various reasons, unwanted. See Hannah Arendt, *The Origins of Totalitarianism* (New York: Harcourt Brace Jovanovich, 1976). Bauman's notion of "waste" is very expressive of this state superfluity. See Zygmunt Bauman, *Wasted Lives: Modernity and its Outcasts* (London: Polity Press, 2004). Cf. Bernard Ogilvie, "Violence et représentation: la production de l'homme jetable", *Lignes*, Vol. 26 (1995), pp. 113–141.

52. Foucault, *Security, Territory, Population, op. cit.*, p. 64.

allowed to take effect are superfluous from the perspective of the goals to be achieved at the massified level.

Perspectival superfluity, however, is necessarily paired with political superfluity—and not only because the impotency or self-limitation of government facing the necessities of natural dynamics ultimately evacuates anything reminiscent of *virtù* from governmental reason.[53] As we have seen above, constructing and rendering governable the economic and the social is conditioned upon forms of knowledge for which the singularity of human action and consequently politics—the appearance of this action—is superfluous as well. Behavioural patterns following natural laws and the homogeneity of public opinion are antithetical to contingency understood—by Arendt—as the correlate of freedom; as the potentiality of "could have been otherwise".[54] Similarly rooted in the prevalence of necessity and the according reconfiguration of contingency and freedom,[55] the interaction of these two forms of superfluity add up to biopolitical abandonment. Subjected to specific governmental rationalities and by way of mobilising, for example, racism or the circulatory imperative of neoliberalism, this biopolitical abandonment can then be understood as the vehicle of the power to disallow life.[56]

Materialising Superfluity

Having outlined above a possible conceptual answer to our original question addressing the paradoxical relationship between a power that aims at the improvement of life and the extent of abandonment it accommodates, in this section the problem is approached from the perspective of its materialisation. In what forms is the biopower to disallow life deployed? How is superfluity crystallised, and how is it being contested? How can we conceptualise resistance to biopolitical abandonment? Pursuing these questions in parallel reflects Foucault's claim about the coexistence of power and resistance and his call to study power from the perspective of its capillary ends.[57] Nevertheless, in order to introduce a possible notion of political resistance into our framework, and due to the analytical potentials of his aesthetic understanding of politics—which shares its Kantian inspirations with many aspects of Foucault's work—below I draw primarily on Rancière's key concepts.[58]

53. "Freedom as inherent in action is perhaps best illustrated by Machiavelli's concept of *virtù*, the excellence with which man answers the opportunities the world opens up before him in the guise of *fortuna*" (Hannah Arendt, "What is Freedom?", in *Between Past and Future: Six Exercises in Political Thought* (New York: Viking Press, 1961), p. 153).

54. John McGowan, *Hannah Arendt: An Introduction* (Minneapolis and London: University of Minnesota Press, 1998), p. 101.

55. See Michael Dillon and Luis Lobo-Guerro, "The Biopolitics of Security in the 21st Century: An Introduction", *Review of International Studies*, Vol. 34, No. 2 (2008), pp. 265–292. For a discussion of the limitations of governmentality theories' usage of contingency see Jacqueline Best, "Ambiguity, Uncertainty, and Risk: Rethinking Indeterminacy", *International Political Sociology*, Vol. 2, No. 4 (2008), pp. 355–374.

56. On the relation of racism and the power to disallow life see Foucault, *"Society Must be Defended"*, *op. cit.*, pp. 239–263. On the "circulatory imperative" see the next section of this paper.

57. See Foucault, *History of Sexuality*, *op. cit.*; *idem*, *"Society Must be Defended"*, *op. cit.*

58. About his understanding of aesthetics that is at the core of politics see Rancière, *Politics of Aesthetics*, *op. cit.*, p. 13.

Central among these is the concept of the distribution of the sensible. As already alluded to through the ideas of the count (which takes account of the community's parts) and the police (that itself is a distribution of the sensible), this notion refers to an ordering of what is perceptible. It is a double system that at once defines what is common and the distinctive parts' shares and positions in relation to that.

> It is a delimitation of spaces and times, of the visible and the invisible, of speech and noise, that simultaneously determines the place and the stakes of politics as a form of experience. Politics [*la politique*] revolves around what is seen and what can be said about it, around who has the ability to see and the talent to speak, around the properties of spaces and the possibilities of times.[59]

The distribution of the sensible thus refers to a "certain framing of time and space" that eventually designates forms of exclusion and inclusion.[60] The delineation of positions in relation to what is common (e.g. the capacity to speak) therefore forms a sensible order, which manifests itself in actual topographies. In a narrower sense—centring on the political community—the police order configures a social hierarchy through allocating places and functions to individuals and groups based on their competencies.

In order to see how this aesthetic approach to governance can bring us closer to conceptualising resistance to biopolitical abandonment, let us consider what a biopolitical distribution of the sensible looks like. If biopolitics primarily means that with modernity the biological life of the human being entered the centre of politics, then what is common is biological life itself. Therefore, the "community" is the species, what is at stake is fostering and disallowing life, and places and functions are allotted based on what counts as the life to be fostered and what can be allowed to die. On these terms, and equipped with Rancière's perspective on the political order as a distribution of places and functions, the crystallisation of superfluity can be traced.

Moreover, this perspective provides the point of departure for thinking superfluity's contestation, as for Rancière the main concern of politics (*le politique*) is "to resist the givenness of a place" and consequently the division based on which that place is allocated.[61] Politics thus triggers the re-partitioning of the police logic, a reordering of what is visible and audible, and of what is regarded as political. This latter aspect explains why it generally and necessarily "occurs 'out-of-place', in a place which was not supposed to be political".[62] However, this heterogeneity is not a sufficient condition for politics to happen, for politics is not a default attribute of a specific place or a particular social group—be they within or without the supposed boundaries of the political—hence, it is not alternative distribution of the sensible opposing the police. Instead, politics is an event: a singular act of creating a stage of appearance. This is exactly what "the part

59. *Ibid.*, p. 12.

60. Rancière, "The Thinking of Dissensus: Politics and Aesthetics", Paper presented at the "Fidelity to the Disagreement: Jacques Rancière and the Political" conference, Goldsmiths College, London, 16–17 September 2003.

61. Mustafa Dikeç, Badlands of the Republic: Space, Politics, and Urban Policy (Malden, MA, Oxford and Victoria: Blackwell, 2007), p. 17.

62. Rancière, "The Thinking of Dissensus", *op. cit.*, p. 4.

that has no part" refers to: those who are uncounted appear so because they cannot be identified with any existing or "real" part of society; they are excess to the count. When they gain visibility as the uncounted people and thus enact the basic miscount, they do so by filling an empty category: the *demos* is, essentially, an empty name. It is filled up with a community of people through the process of subjectification and based on the presumption of equality.

Whereas any logic of the police is the logic of hierarchy, of inequality, politics is the process of equality; it always implements the basic presumption of the equality of any individual being equal with everyone else. The tension from which politics emerges, that is, the fundamental wrong, is exactly this: the heterogeneity of the hierarchical social order and the basic equality of "any speaking being with any other speaking being".[63] A social order is always contingent because it rests on this basic equality:

> There is order in society because some people command and others obey, but in order to obey an order at least two things are required: you must understand the order and you must understand that you must obey it. And to do that, you must already be the equal of the person who is ordering you.[64]

The biopolitical dissolution of this basic equality in the homogeneity of "public opinion" and the "collective interest" works precisely against this "sheer contingency" of social orders. What gives rise to the subjectification of the uncounted if representable barriers are effaced by the triumph of the consensus? As Rancière notes, our time, having renounced the (divisionary) promise of politics, entails a space equally freed of division, that is, "a new configuration of political space, the free development of consensual force adequate to the free and apolitical development of production and circulation".[65]Arguably, (neo-)liberal governmental rationality—guarded by the post-democratic consensus—is a distribution of the sensible that is defined primarily by the necessity of circulation. Thus, the allotment of places and functions reflects above all an account of the capacity or the incapacity to circulate. As such, it crystallises the literal superfluity of globalised capitalism. Let us therefore take a closer look at the circulatory imperative.

The most evident point of departure for this survey is Foucault's differentiation between the territorial rule of the sovereign, the prescriptive rule of disciplinary power and the laisser-faire practices of regulatory biopower.[66] This latter model of power operates primarily through apparatuses of security. That is, through assemblages of practices targeting the givens of the domain to be governed in a way that aims to optimise its beneficial processes and minimise those which are potentially detrimental. As Foucault shows in the example of the town, while security apparatuses continue to deploy technologies of power characteristic of sovereign and disciplinary modes of power, their dominant feature is centred on the notion of freedom; freedom not as an ideology but as a technology of power. Freedom as a technology of power implies that reality is allowed to develop according to its own laws—a technology fundamentally featured in liberalism, as

63. *Idem*, *Dis-agreement*, *op. cit.*, p. 30.
64. *Ibid.*, p. 16.
65. See Rancière, *On the Shores of Politics* (London and New York: Verso, 1995), p. 6.
66. Foucault, *Security, Territory, Population*, *op. cit.*

Foucault notes.[67] In terms of the governance of the developing town, this takes the shape of facilitating circulation. All the functions (hygiene, internal and external trade, and surveillance) that were expected to be secured by the restructuring of the 18th-century town and that were implemented primarily through the construction of roads through and within the town, aimed at "organizing circulation, eliminating its dangerous elements, making a division between good and bad circulation, and maximizing the good circulation by diminishing the bad".[68] Hence the substantial division that, preserved and generalised in neoliberal governmental rationalities, defines contemporary topographies within which the literal superfluity of marginalised people is materialised.

A possible illustration of how this happens, of how such topographies are moulded by different rationalities and technologies of power could be the case of American anti-homeless laws discussed by Don Mitchell. Working within the framework of critical geography, Mitchell's point of departure is the claim that the globalisation of capital is conditioned upon the "production and reproduction of certain *kinds* of spaces" and that "[f]or capital to be free, it must also be fixed in a place".[69] As a result of the constant circulation of capital and the parallel masking of its need for fixity, places face increasing uncertainty, the antidote of which seems to be the almost unprecedented extent of investment in improving their attractiveness. Redefining attractiveness through such discourses as that of the "livable city", several cities in the United States complemented projects of building conference halls and sports centres with what Mitchell terms the "annihilation of space by law": "a legal remedy that seeks to cleanse the streets of those left behind by globalization and other secular changes in the economy by simply erasing the spaces in which they must live".[70] By means of the anti-homeless laws, a strange combination of disciplinary anatomopolitics and sovereign power is put in place: basic physical needs are criminalised through the prohibition of their practice in public places. As it is only homeless people who are compelled to public urination or sleeping in public spaces, Mitchell's conclusion that "these laws attempt not just the annihilation of space, but also the annihilation of the people who live in it" seems to be grounded.[71] Beyond disciplinary and sovereign technologies of power and, perhaps dominating these, we encounter here security apparatuses working to disable bad circulation in order to foster good circulation, and are thus faced with an instance of the circulatory distribution of the sensible.[72]

By pointing to the way in which apparatuses of security instrumentalise law—typically the means of the sovereign model of power—this example draws attention to yet another aspect of the post-democratic order, one that further obstructs

67. *Ibid.*, p. 69.

68. *Ibid.*, p. 34.

69. Don Mitchell, "The Annihilation of Space by Law: The Roots and Implications of Anti-homeless Laws in the United States", *Antipode*, Vol. 29, No. 3 (1997), p. 304; original emphasis.

70. *Ibid.*, p. 305.

71. *Ibid.*

72. Consider, in particular, the passages describing the construction of beggars and other homeless people as impediments to the sufficient extent of consumption: "There is another, perhaps more important, danger posed by those sitting and lying on streets: 'many people see those sitting or lying on the sidewalk and—either because they expect to be solicited or otherwise feel apprehensive—avoid the area. This deters them from shopping at adjacent businesses, contributing to the failure of some and damaging others, costing Seattle jobs and essential tax revenue'" (*ibid.*, p. 309).

the occurrence of politics: the factualisation of law.[73] This phenomenon, complementing the constant reference to the objective necessities deriving from the imperative of circulation, serves as the juridical legitimisation of the "modest state". On the one hand, it refers to the "mimesis of the political practice of litigation" consisting in the minimal government's self-imposition to constitutional control.[74] On the other hand, it implies the equalisation of law and fact in service of circulation.[75] As we will see in the next section, this collision has a significant role in the contemporary management of superfluity.

Superfluous and Uncounted

Arriving finally at the question of resistance, in this section I look at struggles in and around a kind of space that seems to be paradigmatic of our present—possibly more so than the Camp.[76] Generally subject to similar mechanisms of power to those traced in the example of the anti-homeless laws but present on a much larger scale, this sort of space is the shantytown, the *favelas*, the *gecekondu*:[77] located for the most part on the peripheries of mostly "Third World" metropolises and populated by millions of people who cannot afford other forms of accommodation. As I would like to show below, present-day shantytowns that are supposed to be effaced from spectacular visions of urban development might also be paradoxical places when, eventually bouncing into visibility, they are able to disrupt the distribution of the sensible that constructs them as spaces of abandonment. While this claim implies that shantytowns can be seen as *the* political spaces of our present, clearly nothing would be more dissonant with whatever Rancière says about politics than to argue so. Therefore, in what follows, I focus on the particular movement of the South African Abahlali baseMjondolo,[78] for, on occasions, their struggle appears to be able to disrupt the dynamics of biopolitical abandonment and so is able to guide a potential conceptualisation of resistance thereto.[79]

As my point of departure I take the distinction between the population and the people, which Foucault, drawing on Louis-Paul Abeille, introduces when discussing the event of scarcity as touched upon above. According to Abeille, scarcity can

73. Deranty draws a parallel between Rancière's and Jean-Luc Nancy's diagnoses of the phenomenon and refers to it by Nancy's term: "the juridification of the social"; see Deranty, "Rancière's Political Ontology", *op. cit.*, p. 12.

74. Rancière, *Dis-agreement*, *op. cit.*, p. 109.

75. The collision of fact and law that Rancière discusses should not be confused with their indistinction theorised by Agamben in *Homo Sacer*, *op. cit.* (e.g. p. 170). Cf. Rancière, *ibid.*, p. 112; *idem*, "Who is the Subject of the Rights of Man?", *South Atlantic Quarterly*, Vol. 103, Nos. 2–3 (2004), pp. 297–310.

76. The Camp is the "biopolitical paradigm of the modern", according to Agamben in *Homo Sacer*, *op. cit.* (e.g. pp. 9, 123).

77. These are the Brazilian and Turkish words, respectively, for shantytowns. The latter phrase means "it happened at night". "For years, Turkey's squatters built at night to take advantage of an ancient legal precept that said, essentially, that if they started construction at dusk and were moved in by sunrise without being discovered by the authorities, they gained legal standing and could not be evicted without a court fight" (Robert Neuwirth, *Shadow Cities: A Billion Squatters, a New Urban World* (London and New York: Routledge, 2006), p. 8.

78. Literally: "the people who live in the shacks".

79. Beyond available texts of the movement (mostly online at <www.abahlali.org>), I draw on field research carried out with the movement. At the time of writing the research is still ongoing, and the conclusions, therefore, should be regarded as preliminary.

be the chimera it has to be only if "people conduct themselves properly" and so "really act as members of the population".[80] Those, on the other hand, who refuse to conduct themselves properly and disrupt the desirable development of the dynamics of scarcity, consequently even risking revolution, place themselves outside of the population and thus belong to another category: the people—hence the epigraph of this paper.[81] The significance of this distinction lies in its allusion to Rancière's understanding of politics and the people (or, alternatively, the *demos*), as these conceptions posit disruption as essentially political. While not every disruption of the conduct of conduct should be read as at once political,[82] when it is associated with the subjectification of the "people" through processing a fundamental miscount, indeed we encounter politics.

Arguably, the Abahlali's struggle presents such a case. Their resistance is almost fully condensed in this statement: "We are the people who do not count"—as it is primarily by rejecting-through-assuming the status of superfluity that the South African shack-dwellers' movement emerges as a political subject.[83] A political subject conscious of the disruptive effect of presuming equality: "If you want to unite and to make a culture that people should be equal then you are invading the space that is forbidden to you, you are threatening the system."[84] The shack-dwellers' persistent invasion of this forbidden space started with a real invasion: a spontaneous road blockade in Durban in early 2005 as a reaction to rumours about the eThekwini (Metropolitan Durban) Municipality selling to a brick factory a piece of land that was promised to the dwellers of the Clare Estate settlement a few weeks earlier.[85] It was this event, fuelled by anger and the feeling of betrayal due to the Municipality's neglect of the agreement that created the community. "The movement grew out of a spontaneous blockade, of our radical anger and frustration. It was not preceded by intellectual work, but afterwards the movement was formed because we realized that we are not on our own."[86] So emerged the "politics of the poor", or, as it is more often referred to, the "living politics".

In turn, living politics rooted in this process of political subjectification and the notion of life it revolves around renders the Abahlali's struggle able to disturb the processes of biopolitical abandonment.[87] Abahlali contest the forced mobility of

80. Foucault, *Security, Territory, Population*, *op. cit.*, p. 65.

81. *Ibid.*, p. 66.

82. Here again Zanotti's and Best's (*op. cit.*) arguments can be relevant.

83. S'bu Zikode, "Opening Remarks at the Meeting to Build Consensus in Support of Participatory Upgrades in Cities and against Forced Removals to Rural Dumping Grounds", available: <http://abahlali.org/node/3627> (accessed 22 June 2009).

84. Richard Pithouse, "To Resist all Degradations and Divisions—An Interview with S'bu Zikode", *Pambazuka News* (30 April 2009), available: <http://www.pambazuka.org/en/category/comment/55955> (accessed 22 June 2009).

85. Cf. *idem*, "Struggle is a School: The Rise of a Shack Dwellers' Movement in Durban, South Africa", *Monthly Review*, Vol. 57, No. 9 (2005), available: <http://www.monthlyreview.org/0206pithouse.htm> (accessed 22 June 2009).

86. S'bu Zikode, author's notes, 6 May 2009. On the subjectifying force of anger see Simon Critchley, *Infinitely Demanding: Ethics of Commitment, Politics of Resistance* (London: Verso, 2007), p. 130: "[Political] disappointment provokes an experience of injustice and the feeling of anger. I think anger is very important, and, contrary to the classical tradition, in Seneca say, I think it is the first political emotion. It is often anger that moves the subject to action."

87. The most evident manifestation of these processes in present-day South Africa—recalling what was said above about the correlation of the prevalence of the circulative imperative and the

superfluous life by presenting biopolitics with a life heterogeneous to it; by fissuring the homogeneity of species being. It does so primarily through challenging an element central to governing the life of the population: the messianic spatio-temporality of development. To the effects that the circulatory dynamics of development take on the non-pertinent level that incidentally is populated by the shack-dwellers—effects that literally disallow their lives[88]—living politics opposes a manifold insistence on proximity. The appropriation of the idea of *in situ* upgrade—a policy that has been propagated by the UN-Habitat in recent years—for instance, reflects well the motives of this insistence.[89] Relocation to distant housing areas in the unforeseeable future, the no-place of eviction and the no-time of "informality" are opposed to the here-and-now in the demand of developing existing settlements.

This opposition is further reinforced by the insistence on equality and singularity. Both can be reflected through the role of understanding (the ability and the fundamental willingness to understand) within living politics. Understanding is at once the form and the content of living politics—indivisibly rooted in and reaffirming the basic presumption of every human being's equality. On the one hand, one of the most important principles of living politics is that everyone must understand it. With a view to the Abahlali's emphasis on everyone's capability to think and to comprehend, clearly this imperative should not be read as a patronising intention to keep the slogans accessible for an ignorant mass. Quite the contrary: it reflects an ability to understand that rests on a fundamental presumption of equality: "we are all human beings, and so our needs are all, one way or another, similar".[90] Crucially, as the parallel with Rancière's interpretation of Joseph Jacotot's conception of the equality of intellect shows,[91] so conceiving of the capability to understand defies another aspect of the developmental temporality by eliminating the need for explanation:

> I explain an idea to someone because I suppose that he wouldn't understand it if I didn't explain it to him [. . .] Rather than eliminating incapacity, explanation, in fact, creates it. It does this in part by establishing the

factualisation of law—is the series of attempts to change the legislative regulation regarding shack settlements and illegal land occupation. Neutralising the pro-poor elements of earlier legislation, among them the Constitution that famously endorses a wide array of social and economic rights, it now seems that the official state policy towards shack-dwellers—regardless of a growing backlog in the number of low-cost houses built and the number of people entitled to them—is eviction. See, for example, Marie Huchezermeyer, "Comment on KwaZulu-Natal Elimination and Prevention of Reemergence of Slums Bill", in Marie Huchezermeyer and Aly Karam (eds.), *Informal Settlements: A Perpetual Challenge?* (Cape Town: UCT Press, 2006); Centre on Housing Rights and Evictions (COHRE), *Business as Usual? Housing Rights and Slum Eradication in Durban, South Africa* (Geneva: COHRE, 2008), available: <http://www.cohre.org/store/attachments/081007%20Business%20as%20Usual_final.print.pdf> (accessed 22 June 2009), pp. 61, 104.

88. One of the greatest threats is fire: in lacking electricity, people use candles and paraffin stoves, which can cause huge fires in minutes, as the shacks are built mostly of flammable material—and are built very close to each other. In lacking water, too, a candle flipping over can lead to disasters. Cf. Matt Birkinshaw, *A Big Devil in the Jondolos: The Politics of Shack Fires* (Durban: Abahlali baseMjondolo 2008), available: <http://abahlali.org/node/4013> (accessed 22 June 2009).

89. See, for example, the UN-Habitat's Participatory Slum Upgrading Programme at <http://www.unhabitat.org/categories.asp?catid=592> (accessed 22 June 2009).

90. Zikode, quoted in Pithouse, "To Resist all Degradations and Divisions", *op. cit.*

91. See Jacques Rancière, *The Ignorant Schoolmaster* (Stanford: Stanford University Press, 1991).

> temporal structure of delay ("a little further along," "a little later," "a few more explanations and you'll see the light") that, writ large, would become the whole nineteenth century myth of Progress: "the pedagogical fiction erected into the fiction of the whole society," and the general infantilisation of the individuals who compose it.[92]

Singularity, on the other hand, is inherent in living politics' insistence on the willingness to understand: as one of its most important functions, living politics provides a space for everyone's narrative of his or her individual suffering and, in turn, it is exactly these narratives that give it its content.[93] Thus, these narratives of pain and injustice trigger the enactment of *dissensus*. In light of this, as Rancière argues, the separation of the political sphere from the sphere of life necessities makes no sense indeed.[94]

To be sure, the above sketch of Abahlali baseMjondolo's struggle is only the first step in the process of understanding it. In the context of the discussion in this article, however, I did not aim to do more. What this section was intended to point to through outlining the main facets of the South African shack-dwellers' politics is the possibility to think resistance to biopolitical abandonment as it has been conceptualised in the preceding sections. The living politics of Abahlali bears this possibility because through its political subjectification stemming from the presumption of equality and the singularity of human life, apparently, it is able to forge the caesura between the pertinent and the non-pertinent levels of governance posited by modern governmental rationality and is thus able to challenge both the perspectival and the political superfluity inscribed into it. For this, at least, in a world where processes of economic and social marginalisation materialise superfluity to an unprecedented degree, much is to be learned from its struggles.

Conclusion

Starting from the puzzle posed by the ultimate aim of modern governmental rationality, that is, the nurturing of the population and its tendencies to exclude large parts of the same population from the spectrum of its care, this paper argues that abandonment is always already inscribed into this rationality. In contradiction to Agamben, inscribed abandonment is not attributed to sovereign power and its original activity of producing bare life, but is traced back to typically modern processes of transforming the political as problematised by Arendt and Foucault. Complementing these two thinkers' work with those of Hacking and Rancière, it is argued that due to the naturalisation of the political that occurred with the construction of economy as an autonomous field and the forms of knowledge correlative to this construction, and culminating in what is referred to as the literalisation of the count, contingency and plurality of human action has been eliminated from governmental reason. This elimination of contingency, it is claimed, can be understood as the biopolitical abandonment inscribed into modern governmental rationality and, therefore, can be seen as enabling the biopower to disallow life.

92. Kristin Ross, "Rancière and the Practice of Equality", *Social Text*, Vol. 29 (1991), p. 67.
93. S'bu Zikode, author's interview, 2 June 2009.
94. Rancière, "Who is the Subject of the Rights of Man?", *op. cit.*

The aim of conceptualising biopolitical abandonment by drawing on the Arendtian notion of superfluity is to enable an understanding of contemporary forms of extreme social and economic marginalisation and potential forms of resistance to these. Based on the endorsement of Rancière's aesthetic understanding of politics that centres on the notion of the distribution of the sensible, grasping the dynamics of biopolitical abandonment as spatio-temporal articulations is suggested. Still in line with Rancière's thought—which posits politics as the disruption of hierarchical orders of the sensible—and based on a discussion of the South African shack-dwellers' struggle, it is argued that the dynamics of biopolitical abandonment can eventually be diverted—despite the tendency within (global) neoliberal governance to equalise law and fact, so working towards impeding the political subjectification of those who have no part.

Foucault[95] famously claimed that power relations have to be subject to an ascending analysis and that resistance is a diagnostic of power. On the one hand, this implies that the close-up study of specific movements such as the Abahlali baseMjondolo potentially sheds light on the larger dynamics of global neoliberal governance. On the other hand, but still tied to studying instances of political resistance, this claim possibly guides an alternative way of conceptualising communities of our present. As this paper intended to show in contradiction to recent criticisms of the "biopolitical approach",[96] by providing a perspective from which to look at practices and rationalities of power at points where they crystallise and where they are contested, Foucault's thought does serve as a point of departure for thinking the political today. Although it might not suffice in itself to think contemporary forms of political resistance, complemented with an approach to politics as disrupting spatio-temporal orders defined by governmental rationalities, Foucault's concept of biopolitics forcefully grounds criticisms of what is thought to be given. Taking this criticism further through Rancière's understanding of the political as a singular event of resisting the givenness of a place, the *demos* might be found where one would never look for it.

95. See Foucault, *"Society Must be Defended"*, *op. cit.*, p. 30 and *Idem*, "Subject and Power", *op. cit.*, p. 780.

96. See, e.g., David Chandler, "Critiquing Liberal Cosmopolitanism? The Limits of the Biopolitical Approach", *International Political Sociology*, Vol. 3, No. 1 (2009), pp. 53–70; and Jan Selby, "Engaging Foucault: Discourse, Liberal Governance, and the Limits of Foucauldian IR", *International Relations*, Vol. 21, No. 3 (2007), pp. 324–345.

Rethinking Foucault in International Relations: Promiscuity and Unfaithfulness

ANDREW W. NEAL

Foucault and those who employ his work are coming under two sets of demands. First, what can Foucault say about the international and about capitalism? Second, are claims about "global governmentality", which upscale Foucauldian concepts, defensible? The essays in this special issue do a good job of meeting both demands. In this short response, I will try to give a very general answer explaining why.

Foucault studied rationalities of government, that is, the emergent self-understandings of those who would seek to govern, or through their ideas, innovations, practices and interactions, contribute to assemblages that amount to forms of government. What seemed to intrigue Foucault was how new or altered rationalities of government emerge in history and take hold. These are not new ideologies, eras, paradigms or systems that take a total hold on individuals, groups and populations, but practices that cohere just enough to become significant and describable.

Foucault does not describe these assemblages just out of historical interest, although to a casual reader this might at first appear to be the case. His meticulous descriptive histories have a significance; this is where the term "history of the present" has taken hold. But this significance is not always obvious and is often rather oblique. For example, *The Birth of Biopolitics* is the latest of the Collège de France lecture series to be published in English and to be received enthusiastically. It speaks to the present by helping us to understand the rationalities that animate "neoliberal" forms of government and make certain governmental practices make sense and appear intelligible.

Does this mean that through Foucault we can find the origins of neoliberalism? Perhaps, perhaps not. Origins are never simple starting points, carte blanche. Ideas and practices build on previous ideas and practices, alter them, amend them, improve them, add to them. Foucault's genealogical problematisation of origins is well understood. From this perspective neoliberalism is a rationality of government that appears rational in specific historical circumstances, under certain historical conditions. Foucault describes the emergence of neoliberalism as a rationality of government, but does this mean that neoliberalism is now with us and we can now treat it as an object? Does it now *exist*?

The danger is that we take lessons from Foucault and then forget them, or at least incorporate them into our academic systems of thought. The first time we encounter Foucault we find histories that reveal so much about the present, and concepts that capture and articulate so much of contemporary political experience. But perhaps the real lessons come only when we *return* to Foucault, maybe after working on a different project or studying another thinker. When we return to Foucault, he forces us to reassess our concepts, even if we learned those concepts from Foucault in the first place. Even if we feel we have learned what neoliberalism is, for example, we return to Foucault only to be reminded that *it does not exist*. It is only a problem around which "different events and practices . . . are apparently organized".[1]

When we return to Foucault, we realise that one part of Foucault's oeuvre does not support what we learned in another part. It is not that Foucault's work suffers from internal contradictions. The mistake is ours not his. When we read *Discipline and Punish*, for example, we take the lesson that disciplinary power plays an important role in both historical and contemporary society, through institutions and practices that form individuals in a manner appropriate to the defining problems of the time or locale. When we return to Foucault elsewhere, we then find that as soon as disciplinary power appeared, it was already disappearing, not completely, but into the background. Foucault then describes a new problematisation, perhaps neoliberalism, perhaps biopolitics, perhaps governmentality. It is not so much that these new problematisations replace the old ones in historical series. It is that these new problematisations replace the position of the old ones in our academic systems of thought. Where we saw disciplinary power permeating the present, we now see the new concept eclipsing the old. And every time we return to Foucault we are forced to adjust the Foucauldian frames of reference and systems of thought that we had become accustomed to working with.

If we treat Foucault as a historian, we will always be able to pick holes. His work is notoriously lacking in references and difficult to "verify". While Foucault's prose does give the impression of historical fact, it must always be remembered that this is not entirely the point of his work. His aim is not that of the Cambridge School of history, of "seeing things their way".[2] His histories speak to us in a way that other histories do not. The difference is that Foucault is not a historian of record or fact, but nor is he a theoretical historian, imposing systematisation on the raw matter of history, perhaps in the form of stages of capitalism or development. He describes histories at the surface level which can only ever be incomplete. He looks for marginal figures, or documents that might be unfamiliar, and gives them an importance that others would not. He does not impose a system that makes sense of history as a whole, or impose unity and meaning on an otherwise diverse group of statements and objects. Rather, he looks for emergent systems of rules that are specific to particular historical discourses. This is the "law of what can be said",[3] the set of historical, discursive, epistemological and ontological conditions that allow statements to appear meaningful or necessary. But these conditions are only ever historically specific and localised. Remember that the title Foucault created for his chair at the Collège de France was "The History of Systems of Thought".

1. Michel Foucault, *The Birth of Biopolitics: Lectures at the Collège de France, 1978–1979* (ed. Arnold Davidson, trans. Graham Burchell) (Basingstoke: Palgrave Macmillan, 2008), p. 3.

2. Quentin Skinner, *Visions of Politics*: Vol. 1, *Regarding Method* (Cambridge: Cambridge University Press, 2002).

3. Michel Foucault, *The Archaeology of Knowledge* (London: Routledge, 1989 (2002 printing)), pp. 145–146.

The difficulty with trying to make Foucault speak to questions of capital or the international is that there are strong disciplinary principles and authorities in play that wish to give primary authority to those concepts: that somehow relations of capital or differential international development are the root sources of politics or injustice, or that the international is the fundamental condition of possibility for relations of antagonism or war. But the most important lesson from Foucault is that no concept can be sacred. Every concept has a history, not in the sense that its effects can be detected across history, but in the sense that every concept emerged through discursive relations mobilised around specific historical problematisations. Indeed, every history, including that of capital and the international, has a history.

Foucault is a writer of histories who is not really a historian, and a political thinker who does not really give us a clear articulation of politics. Foucault tells us where we might find politics, or at least find things that have political significance. He describes instances where things became problems for those concerned with government in the broadest sense. But while Foucault suggests where we might find politics, he does not give us an obvious politics. He gives us no maxims, no prescriptions, no codes to follow, no positions to take. He does gesture towards notions such as resistance, and, later on, ethical self-fashioning, but again these are things he describes within specific historical systems of thought.

There is some irony in taking this historian who is not a historian and political theorist who is not a political theorist and trying to rethink him for International Relations (IR) or for the study of relations of capital. It is not possible to do so and remain faithful to those disciplines. To engage with the highly promiscuous thinker that is Foucault is to be unfaithful. It is not possible to engage with Foucault while holding onto even the most rudimentary of disciplinary commitments, such as the idea that "the international" is an object of study. For Foucault, such objects do not exist, they are only problems around which "different events and practices ... are apparently organized".[4]

Engaging with Foucault should also mean being unfaithful to oneself. When we return to Foucault and find him reflecting on his own practice, perhaps at the beginning of a new year of lectures at the Collège de France, he shows no compunction in dismissing his previous work, expressing his dissatisfaction with it, and articulating a new problem that will now animate his work instead. We should do the same, especially if that means being unfaithful to Foucault himself. We should not allow Foucauldian concepts to become disciplinary when Foucault did not think twice about abandoning them. In *The Birth of Biopolitics*, Foucault explains that his choice of method was

> obviously and explicitly a way of not taking as a primary, original, and already given object, notions such as the sovereign, sovereignty, the people, subjects, the state, and civil society, that is to say, all those universals employed by sociological analysis, historical analysis, and political philosophy in order to account for real governmental practice.[5]

4. *Idem, The Birth of Biopolitics, op. cit.*, p. 3.
5. *Ibid.*

We should make the same decision when it comes to concepts such as governmentality. That is not a prohibition on using the term, but a warning against allowing it to become an *object*. The essays in this special issue convey a similar message: Foucault does not allow us to give ontological or epistemological priority to any object, concept or category, including "the international" and "capitalism". Foucauldian concepts such as "governmentality" can only be used to problematise diverse techniques of power, not to systematise, especially not on a global scale. While the essays learn from Foucault the lesson of *being critical and not dogmatic*, they are also prepared to be unfaithful to him, to challenge him, and to use him in creative ways.

Recently, a new unabridged version of Foucault's *History of Madness* was published in English. This has not attracted a great deal of attention from IR scholars, who are more engaged by the ongoing publication of the Collège de France lectures. This is a shame, for although *History of Madness* does not speak so directly to contemporary concerns about governmentality, biopolitics, sovereignty and war it does tell us a great deal about Foucault's method. We know very well the famous line that "Madness does not exist, but that does not mean it is nothing", and we also know that Foucault did not actually write that, but elsewhere confirmed that it could indeed "be written".[6] The point is that madness is not an object, but nevertheless it came to be treated as one in many different contexts, animating orbiting knowledges, discourses, moral economies, institutions, subjectivities and techniques.

Looking more closely at the full manuscript of *History of Madness*, above the endless detail, we can start to see how Foucault approaches this thing that "does not exist"; not as some kind of fiction, but as a haunting presence at the heart of Western cultures for centuries. It is not a singular presence; in fact it is never present for more than a fleeting moment, even though it occupies many hundreds of pages. Foucault describes this thing that does not exist as a constantly moving target, something that never exists for long enough to be described definitively. If we look closely at the way Foucault writes about madness, we can see that as soon as he mentions a new discourse it is already disappearing. This constant disappearance is expressed not only in the imagery of the text but also in the movement of the writing itself. Somehow, every description is annulled within a few pages. As soon as a seemingly decisive discourse of madness appears, it is already disappearing.

In the chapter on "The Proper Use of Liberty", Foucault describes a shift from a negative understanding of madness in the classical age, in which madness is a kind of negative subjectivity, to a positive understanding in the late 18th century in which madness becomes an object of knowledge: "Madness was no longer to be inscribed in the negativity of existence, as one of its most brutal figures, but now progressively took its place in the positivity of known things."[7] On its own, this might be interpreted as a kind of paradigm shift or developmental stage, except that similar displacements of one understanding of madness by another occur every few pages. Looking at the first chapter we can see this constant movement, every description already a displacement of something else:

6. Paul Veyne, "Foucault Revolutionizes History", in Arnold I. Davidson (ed.), *Foucault and his Interlocutors* (Chicago and London: University of Chicago Press, 1997), p. 170.

7. Michel Foucault, *History of Madness* (trans. Jean Khalfa and Jonathan Murphy) (London and New York: Routledge, 2006), p. 443.

first "the strange disappearance of leprosy",[8] then in the Middle Ages "Madness now has an absolute privilege",[9] but then at the dawn of the classical age, "Madness has ceased to be an eschatological figure at the edge of the world of mankind and of death."[10] It is difficult to convey here that in the text there is no duration, no progression, no fixity. It would not be possible to place these different discourses of madness on a timeline or diagram, for although they disappear sequentially they continually return to haunt other understandings of madness.

In Foucault's work we confront particular practices of power and historically specific systems of thought. Foucault does not offer a generalisable theory or system of thought, and to try to incorporate him into a systemic way of thinking that privileges certain categories will never work. Why should a thinker who denies and subverts disciplinary boundaries be disciplined by a discipline? Foucault refused to discipline his thought, turning on his own insights within a few pages or from one lecture series to the next. Nothing is sacred in Foucault. This movement within his work sets up creative harmonics of addition and subtraction. It makes little sense to ask what Foucault can do for IR, when his critical achievement is to help make IR a less disciplined discipline.

8. *Ibid.*, p. 5.
9. *Ibid.*, p. 22.
10. *Ibid.*, p. 41.

Index